McGraw-Hill
Mathematics

Daily Homework
Practice

4

McGraw-Hill
School Division

New York   Farmington

McGraw-Hill School Division

A Division of The McGraw·Hill Companies

Copyright © McGraw-Hill School Division,
a Division of the Educational and Professional Publishing Group of The McGraw-Hill Companies, Inc.
All rights reserved.

McGraw-Hill School Division
Two Penn Plaza
New York, New York 10121-2298

Printed in the United States of America

ISBN 0-02-100286-X / 4

4 5 6 7 8 9 024 05 04 03 02 01

GRADE 4
Contents

Chapter 4: Multiplication and Division Facts

Chapter 5: Multiply by 1-Digit Numbers

Chapter 6: Multiply by 2-Digit Numbers

Chapter 7: Divide by 1-Digit Numbers

Chapter 8: Divide by 2-Digit Numbers

Chapter 9: Measurement

Chapter 10: Geometry

Chapter 11: Fractions and Probability

Chapter 12: Fraction Operations

Chapter 13: Relate Fractions and Decimals

Chapter 14: Decimal Operations

1-1 ▶ Explore: How Big Is a Million?

Graph it.

Take a piece of graph paper and mark off a 10 × 10 grid as shown below.

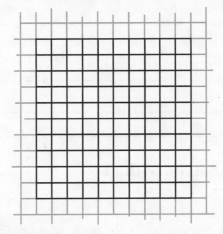

1. How big an area have you marked off? Show your calculation.

_____

Now mark off nine more of these grids. Cut out all of your grids and line them up in a row.

2. How big an area do you have now? Show your calculation.

_____

Now make nine more rows like this and arrange them in a large square.

3. How big an area do you have now? Show your calculation.

_____

4. To make a million, how many squares the size of this large square do you have to make? _____

5. Write a million in figures, then using 10s and ×s.

_____

Spiral Review

Write the number that is missing in each pattern.

6. 54, _____, 56

7. 22, 24, 26, 28, _____

8. 7, 19, 31, _____

9. 15, _____, 60, 120

10. 2, 4, 7, 11, 16, _____

11. 80, 60, _____, 20, 0

Name_____

 **I·2** **Place Value Through Millions**

Write the word name and the expanded form for each number.

1. 4,002 _____

2. 70,506 _____

3. 4,900,015 _____

   _____

4. 300,056,902 _____

   _____

Write the value of each underlined digit.

5. 4<u>1</u>5,338 _____   6. 3<u>4</u>5,126,120 _____

7. 7<u>6</u>,500,039 _____   8. 676,502,1<u>5</u>2 _____

Write each number in standard form.

9. seventy-nine million, four hundred thousand, two _____

10. eight hundred seventy-three thousand, four hundred sixteen _____

11. 30,000 + 4,000 + 90 + 8 _____

12. 7,000,000,000 + 40,000 + 500 + 60 + 6 _____

Problem Solving

Use data from the chart for problems 13–14.

13. Park officials estimate they will serve 21,000 more meals in 2001 than in 2000. How many will this be? _____

14. What number is in the ten thousands place of the 1998 amount? _____

MEALS SERVED AT THE WILD ANIMAL PARK	
1998	2,250,000
1999	2,401,000
2000	2,418,000

Spiral Review

15. 339 + 142 = _____   16. 9 × 7 = _____

17. 976 − 428 = _____   18. 27 ÷ 9 = _____

Name_____

 **1·3** **Compare and Order Numbers and Money**

Compare. Write >, <, or =.

1. 88,787 _____ 88,878

2. $8.99 _____ $9.89

3. $443.86 _____ $443.68

4. 221,122 _____ 221,122

5. $946.60 _____ $94.60

6. 78,543 _____ 617,445

Order from greatest to least.

7. 5,687; 5,499; 5,944 _____

8. $156.61, $161.56; $166.51 _____

Order from least to greatest.

9. 13,551; 135,551; 131,510 _____

10. $842.55; $84.52; $854.42 _____

Problem Solving

11. The heaviest sea mammals, blue whales, can weigh 236,000 pounds. The heaviest land mammals, African bush elephants, can weigh 23,600 pounds. The heaviest dinosaurs weighed about 176,000 pounds. Which is the heaviest animal of all time? Explain your answer.

_____

12. There are about 24,000 species of fish. There are more than 4,000 species of land mammals. There are about 73,000 species of spiders, scorpions, and mites. There are 9,000 species of birds. Which of the groups of animals named has fewer species than birds? Explain your answer.

_____

Spiral Review

13. How many thousands are in ten million? _____

14. What is the value of 8 in 98,566? _____

Name_____

Problem Solving: Reading for Math

Using the Four-Step Process

Use the four-step process to solve.

1. A kangaroo can run 40 miles an hour over short distances. A giraffe can run 37 miles an hour. Over short distances, a brown hare can run 13 miles per hour faster than a giraffe. How fast can a brown hare run? How much faster is this than a kangaroo?

_____

Use data from the table for problems 2–3.

GREATEST WEIGHTS AND LENGTHS
OF AUSTRALIAN ANIMALS

Animal	Weight of Heaviest Male (pounds)	Greatest Head and Body Length (inches)
gray-headed fruit bat	2.6	11.7
kangaroo	200	65
koala	26 (average)	33
platypus	4.8	21
rufous rat kangaroo	7.5	15.5

2. Order the rufous rat kangaroo, the platypus, and the gray-headed fruit bat from shortest to longest. _____

3. Which animal is longer than but weighs less than what other animal?

_____

Spiral Review

4. Order from greatest to least: 567,173; 576,771; 575,673

_____

5. Order from least to greatest: $669.69, $969.99, $696.96

_____

Name_____

1·5 Round Numbers and Money

Round to the given place.

1. 579 to the nearest ten

   _____

2. 35,488 to the nearest thousand

   _____

3. 49,052 to the nearest thousand

   _____

4. 416,508 to the nearest thousand

   _____

5. 416,508 to the nearest ten thousand

   _____

6. 416,508 to the nearest hundred

   thousand _____

7. $56.20 to the nearest dollar

   _____

8. $56.20 to the nearest hundred

   dollars _____

9. $8.55 to the nearest ten cents

   _____

10. $8.55 to the nearest dollar

   _____

Problem Solving

Use this data to solve problems 11–15.

In 1995, the number of Labrador retrievers registered with the American Kennel Club was 132,051. The number of golden retrievers was 64,107. The number of Chesapeake Bay retrievers was 5,069. And the number of Newfoundlands was 2,862.

11. To the nearest ten, how many Newfoundlands were there? _____

12. To the nearest hundred, how many Labradors were there? _____

13. To the nearest ten thousand, how many Labradors were there? _____

Spiral Review

Compare. Write >, <, or =.

14. The number of beagles registered was 57,063. How does this compare with the number of golden retrievers? _____

15. The number of Dalmatians was 36,714. How does this compare with the number of Newfoundlands? _____

Name_____

1·6 Problem Solving: Strategy
Make a Table

At Sheila's Seashell Shop, Sandy picked out and bought the following shells (with the areas where they are from):

giant keyhole limpet—California
West Indian top—Caribbean
chestnut turban—Caribbean
giant button top—Japan
spiked limpet—South Africa
chocolate-lined top—Caribbean

long-spined star—Caribbean
triumphant star—Japan
horned turban—Japan
South African turban—South Africa
West Indian fighting conch—Caribbean
emperor's slit shell—Japan

Use data from the list for problems 1–3. Make a table to solve.

Kind of shell	California	Caribbean	Japan	South Africa

1. How many varieties of star shells did Sandy pick? _____

2. How many shells from Japan did Sandy pick? _____

3. From what place did Sandy pick the most shells? How many? _____

Spiral Review

4. Last month Sheila spent $785.32 on new stock. Round this amount to the nearest dollar. _____

5. Last month Sheila spent $600 for rent and $102.50 for utilities. Round the amount of money she spent for rent and utilities together to the nearest dollar. _____

Name_____

 **I·7** ▶ **Count Money and Make Change**

Write the amount of money shown.

1.

_____

2.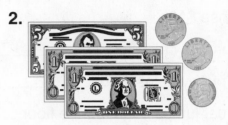

_____

Tell which coins and bills make the amount.

3. $10.24 _____

4. $9.08 _____

5. $1.73 _____

Find the amount of change.

6. Price: $2.28
Amount given: $10.00 _____

7. Price: $7.89
Amount given: $10.00 _____

Problem Solving

Use the data that follows for problems 8–9.

The fishing boat *Jenny S.* caught 1,812 mackerel, 989 codfish, and 1,920 haddock.

8. Did the catch include a greater number of mackerel or haddock? _____

9. About how many codfish were caught, to the nearest hundred? _____

Spiral Review

10. Order these amounts from the least to the greatest: $5.66, $5.97, $5.87

_____

11. Order these amounts from the greatest to the least: $10.47, $9.96, $10.74

_____

Name_____

 **I·8** ▶ **Negative Numbers**

Write a positive or a negative number to represent each situation.

1. You lose $10. _____

2. You grow 4 cm. _____

3. Your puppy gains 4 ounces. _____

4. Your cat jumps 3 feet off the floor. _____

5. A polar bear doesn't mind temperatures 20°F below zero. _____

6. The dolphin dives 5 m below the surface of the pool. _____

7. It is too cold for a turtle at 30°F above zero. _____

Compare. Write < or >. You may use a number line to help.

8. 4 ____ −4   9. 9 ____ −6   10. −6 ____ −9   11. 5 ____ −4

12. 0 ____ 7   13. 0 ____ −7   14. −7 ____ −2   15. −8 ____ −1

Problem Solving

16. One kind of fish has been found in the Atlantic Ocean 8,000 m deep. Write this as a negative number.

    _____

17. The Siberian salamander lives in northern Russia, where winter temperatures may go as low as 56°C below zero. Write this as a negative number.

    _____

Spiral Review

Find the amount of change.

18. With tax, a cat toy costs $3.07. You give the clerk $5.00. _____

19. A large popcorn at the zoo costs $2.75. You give the clerk $10.00. _____

20. With tax, a toy stuffed leopard costs $8.68. You give the clerk $10.00.

    _____

Name_____

2·1 ▶ **Use Properties of Addition**

Complete each set of related number sentences.

1. $5 + v = 12$ _____
  $f + 7 = 12$ _____
  $12 - 7 = f$ _____
  $12 - f = 7$ _____

2. $6 + 15 = t$ _____
  $15 + x = 21$ _____
  $21 - f = 6$ _____
  $21 - 15 = x$ _____

3. $h + 22 = 25$ _____
  $3 + 22 = w$ _____
  $25 - 22 = h$ _____
  $w - 22 = 3$ _____

Find the sum or the difference. Write the related number sentences.

4. $29 - 9$

_____
_____
_____
_____

5. $43 - 0$

_____
_____
_____
_____

6. $11 + 5$

_____
_____
_____
_____

Write the related number sentences for each set of numbers.

7. 4, 10, 14

_____
_____
_____
_____

8. 14, 14, 28

_____
_____
_____

9. 25, 29, 54

_____
_____
_____
_____

Problem Solving

Use data from the table for problems 10–11.

10. How many more Happy Dolls does Annette
  have than Karen? _____

11. What related number sentences could you
  write for the numbers in the chart?

_____ _____ _____ _____

Number of Happy Dolls Owned	
Darlene	17
Annette	30
Karen	13

Spiral Review

12. $7 + 8 =$ _____

13. $9 + 4 =$ _____

14. $11 - 0 =$ _____

Name_____

 **2-2** ▶ **Addition Patterns**

Complete the pattern.

1. 9 + 7 = a _____

    90 + 70 = b _____

    900 + 700 = c _____

    9,000 + 7,000 = d _____

    90,000 + 70,000 = e _____

    900,000 + 700,000 = f _____

2. 8 + 4 = g _____

    80 + 40 = h _____

    800 + 400 = j _____

    8,000 + 4,000 = k _____

    80,000 + 40,000 = m _____

    800,000 + 400,000 = n _____

Add mentally.

3. 400 + 600 _____

4. 5,000 + 9,000 _____

5. 20,000 + 90,000 _____

6. 200,000 + 600,000 _____

7. 900,000 + 100,000 _____

8. 500,000 + 800,000 _____

9. 80,000 + 80,000 _____

10. 8,000 + 8,000 _____

Problem Solving

11. In a game called Make a Million, Alberta had $40,000 in play money and her teammate Grace had $70,000. How much did they have together?

    _____

12. Lewis Publishing printed 500,000 books in 1998 and 600,000 books in 1999. How many books did they print in the two years?

    _____

Spiral Review

Round to the nearest thousand.

13. 5,502 _____

14. 76,943 _____

15. 6,249 _____

Round to the nearest ten thousand.

16. 66,400 _____

17. 65,022 _____

18. 84,566 _____

2·3 Add Whole Numbers and Money

Find the sum.

1. 823
  + 917

2. $115.90
  + 76.88

3. 113,596
  104,533
  + 99,811

4. $1,988.52
  765.90
  + 33.37

5. 512 + 663 = _____

6. 889,787 + 263 = _____

7. 11,111 + 286 + 55 = _____

8. 92,096 + 3,380 + 457 = _____

Problem Solving

9. In three pinball games, Ilya scores 566 points, 292 points, and 189 points. What was his total for the three games? _____

10. Wendy played three games of Velcro darts and scored 112, 68, and 56 points. What was her total for the three games? _____

Use data from the table to solve problems 11–12.

11. In the next spelling bee, Greg earns 2,103 points. Between which two players will he rank?

_____

12. Jinnie earns 2,200 points. Which players have more points than she does? _____

Top Fourth-Grade Spellers

Name	Points
Magnus	2,590
Aliki	2,286
Neal	2,190
LaToya	2,154
Ward	2,097

Spiral Review

Compare. Write >, <, or =.

13. 4,599 _____ 5,489

14. 635,090 _____ 660,935

15. 568,924 _____ 559,995

16. 671,667 _____ 671,667

Name_____

 **2·4** **Use Mental Math to Add**

Add mentally.

1. 88 + 46 = _____

2. 34 + 78 = _____

3. 92 + 86 = _____

4. 115 + 198 = _____

5. 156 + 376 = _____

6. 468 + 439 = _____

7. 715 + 196 = _____

8. 323 + 516 = _____

9. 666 + 292 = _____

10. 613 + 392 = _____

11. $445 + $432 = _____

12. $445 + $492 = _____

13. $807 + $452 = _____

14. 1,990 + 1,801 = _____

15. 3,451 + 1,998 = _____

16. 3,670 + 3,550 = _____

17. $1,516 + $1,212 = _____

18. $1,223 + $3,445 = _____

19. $2,780 + $6,001 = _____

20. $4,503 + $2,599 = _____

Problem Solving

21. Kirk spent $5.67 at the grocery store and $2.95 at the newsstand. How much did he spend in all? _____

22. Regina bowled two games. In the first she bowled 114, and in the second she bowled 99. What was her total for the two games? _____

23. Alicia collected 516 tickets at the school fair yesterday. Today she collected 668. How many did she collect on both days? _____

Spiral Review

Find the amount of change.

24. Cost: $0.56
    Amount given: $5.00

    _____

25. Cost: $2.94
    Amount given: $3.00

    _____

Name _____

2·5 Estimate Sums

Estimate each sum.

1. 472 + 518 _____

2. 508 + 91 _____

3. 725 + 390 _____

4. $4.99 + $5.99 _____

5. $4.76 + $4.82 _____

6. $0.93 + $0.27 _____

7. $0.32 + $0.37 _____

8. 4,997 + 366 _____

9. 5,150 + 2,003 _____

10. 506,199 + 565,920 _____

Add. Estimate to check that each answer is reasonable.

11. 2,450 + 289 = _____

12. 45,207 + 5,669 = _____

13. 135,998 + 457,722 = _____

14. 168,900 + 577,281 = _____

Compare. Write > or < to make a true sentence.

15. 155 + 349 _____ 400

16. 557 + 422 _____ 1,000

Problem Solving

17. John has 1,569 bottle caps in his collection. Reynaldo has 1,039. About how many do they have together? _____

18. Dolores gave John all her bottle caps before she moved away. If she gave him 1,409, about how many does he have now? _____

19. Molly's brother, Edward, gave her his bottle cap collection before he went to college. She already had 2,730. He gave her 4,779. About how many does she have now? _____

Spiral Review

Write >, <, or =.

20. 567 _____ 565

21. 4,989 _____ 4,898

22. 156,987 _____ 156,987

23. 1,156,784 _____ 1,165,787

2·6 **Problem Solving: Reading for Math**

Estimate or Exact Answer

Solve. Explain why you gave an estimate or an exact answer.

1. Yoko makes pillows decorated with buttons. Yesterday she sold two, one for $15.50 and the other for $16.75. How much money did she make?

   _____

2. Yoko had 5,023 buttons, and then her grandmother gave her 1,299 more. Does she have more than 7,000 buttons?

   _____

3. Victor found two books of trading stamps that had belonged to his great-grandfather. One book was full and had 1,250 stamps. The other one had only 899 stamps. How many stamps had Victor's great-grandfather saved?

   _____

Use data from the table for problems 4–6.

4. The Astronomy Club is trying to raise money to buy a new telescope. The telescope costs $650. If the club has a holiday card and gift sale, will they make enough money?

   _____

Astronomy Club's Fund-Raising Ideas	
Idea	**Likely Earnings**
Car wash	$275
Dog wash	$125
Cupcake sale	$225
Holiday card sale	$175
Holiday gift sale	$425

5. If the club wants to hold only two events, which combinations will give them at least $650?

   _____

6. About how much money could the club make if it held all five events?

   _____

Spiral Review

7. 145,699 + 78,026 = _____

8. 145,699 + 182,989 = _____

9. 333,807 + 425,661 = _____

10. 598,997 + 51,642 = _____

Name_____

 **2·7** **Subtraction Patterns**

Complete.

1. $17 - 9 = p$ _____
   $170 - 90 = q$ _____
   $1,700 - 900 = r$ _____
   $17,000 - 9,000 = s$ _____
   $170,000 - 90,000 = t$ _____
   $1,700,000 - 900,000 = u$ _____

2. $11 - 6 = v$ _____
   $110 - 60 = w$ _____
   $1,100 - 600 = x$ _____
   $11,000 - 6,000 = y$ _____
   $110,000 - 60,000 = z$ _____
   $1,100,000 - 600,000 = a$ _____

Subtract mentally.

3. $1,900 - 1,000 =$ _____
4. $3,000 - 1,300 =$ _____
5. $5,100 - 2,600 =$ _____
6. $67,000 - 49,000 =$ _____
7. $111,000 - 55,000 =$ _____
8. $190,000 - 100,000 =$ _____
9. $70,000 - 6,000 =$ _____
10. $150,000 - 6,000 =$ _____

Problem Solving

11. Kenji ordered nails and tacks for his scout troop. He ordered 3,000 nails and 1,800 tacks. How many more nails than tacks did he order? _____

12. Malcolm ordered mosaic tiles for the troop. He ordered 30,000 in white and 48,000 in mixed colors. How many more colored tiles than white tiles did he order? _____

Spiral Review

Find the sum or the difference. Write the related number sentences.

13. $5 + 7$

_____

_____

_____

14. $8 + 12$

_____

_____

_____

15. $14 - 8$

_____

_____

_____

16. $10 + 13$

_____

_____

_____

17. $36 - 25$

_____

_____

_____

18. $44 - 18$

_____

_____

_____

Name_____

Subtract.

1. 599 – 426	**2.** 646 – 111	**3.** 716 – 292
4. 596 – 575	**5.** 778 – 666	**6.** 778 – 497
7. 505 – 327	**8.** 671 – 590	**9.** 338 – 209

10. 695 – 299 = _____   **11.** 992 – 808 = _____

12. 564 – 495 = _____   **13.** 408 – 330 = _____

14. 809 – 220 = _____   **15.** 340 – 216 = _____

Solve.

16. The PTA sold a total of 628 tickets for the two performances of the Spring Concert. If 376 of the tickets were for the Friday performance, how many tickets were for the Saturday performance? _____

17. On a tax-free shopping day, Jim bought a new shirt for $12.98. If he gave the clerk a twenty-dollar bill, how much change should he get? _____

Spiral Review

18. 156 + 145 + 82 = _____   **19.** 189 + 156 + 120 = _____

20. 387 + 240 + 122 = _____   **21.** 766 + 230 + 34 = _____

22. 500 + 290 + 187 = _____   **23.** 826 + 498 + 104 = _____

Name _____

2·9 Subtract Whole Numbers and Money

Subtract. Check by adding.

1. 874
    − 443

2. $9.78
    − 5.98

3. 8,718
    − 660

4. 19,890
    − 11,955

5. $51.61
    − 17.62

6. 88,997
    − 53,558

7. 115,336
    − 92,940

8. $555.12
    − 337.46

9. 619,014
    − 323,440

10. $724.49 − $330.83 = _____

11. 451,788 − 245,566 = _____

12. 583,900 − 404,667 = _____

13. $667.89 − $350.04 = _____

Problem Solving

14. Elaine paid $559.98 for a stereo system at a department store. Susan bought the same system on sale at an electronics store for $469.99. How much did Susan save? _____

15. Very Vinyl has 20,063 records for sale. Old Time Disk has 4,566 fewer. How many records does Old Time Disk have for sale? _____

Spiral Review

16. 5,770 + 3,394 = _____

17. 26,546 + 23,950 = _____

18. 32,871 + 45,906 = _____

19. 5,006 + 2,129 = _____

20. 45,199 + 10,760 = _____

Name_____

 **2·10 Regroup Across Zeros**

Subtract. Check by adding.

1. 8,007
    − 4,438

2. $9.00
    − 5.98

3. 8,001
    − 6,666

4. 10,090
    − 7,955

5. $510.06
    − 17.62

6. 800,997
    − 533,558

7. 100,000
    − 92,940

8. $5,005.12
    − 337.46

9. 600,004
    − 399,445

10. $700.24 − $303.88 = _____

11. 401,008 − 245,999 = _____

12. 500,900 − 404,667 = _____

13. $1,000.67 − $357.77 = _____

14. $1,500.06 − $994.57 = _____

Problem Solving

15. In 2000, Doris saw a quilt in the museum that was made in 1804. How old

    was it? _____

16. The Quilt Quorner has 700,000 quilting squares for sale. If Justine buys

    1,152 of them, how many are left? _____

Spiral Review

Estimate.

17. 577,025 + 339,004 _____

18. 265,465 + 239,500 _____

19. 328,711 + 459,066 _____

20. 451,990 + 107,603 _____

Name_____

2·11 **Problem Solving: Strategy**

Write a Number Sentence

Write a number sentence to solve.

1. Consuelo has decided to sell her silkscreen paintings. If supplies cost her $5 for each painting, how much does she have to charge to make $25 profit on each painting? _____

2. Wyatt makes paintings with bits of gravel and glue. If supplies cost him $4.25 for each painting, how much profit will he make if he charges $27 for each painting? _____

3. Melanie spends $1.98 apiece for glue for her junk sculptures. How much less does she spend for supplies than Consuelo? _____

Mixed Strategy Review

Solve. Choose any strategy.

4. Gray wants to earn enough money from selling her ceramics to buy a gift for her grandmother. She sold three pots for $6.25 each. The gift costs $19 plus tax. Does Gray have enough money? _____

Use data from the table for problems 5–6.

5. Blaise wants to buy a tube of each color paint. How much will he have to spend?

_____

6. Keisha has $16. Does she have enough money to buy a tube each of crimson, titanium white, ebony black, and chrome yellow? _____

Cost of Tubes of Acrylic Paint

Crimson	$3.98
Cobalt blue	$3.39
Titanium white	$3.29
Ebony black	$3.29
Burnt sienna	$3.49
Chrome yellow	$4.19
Spring green	$3.79

Spiral Review

Subtract.

7. $560.02 − $351.90 = _____

8. 100,000 − 39,596 = _____

9. 9,004 − 4,731 = _____

10. 70,021 − 55,430 = _____

© McGraw-Hill School Division

Name_____

 **2·12** **Subtract Using Mental Math**

Subtract mentally.

1. 44 – 31 = _____

2. 67 – 9 = _____

3. 91 – 23 = _____

4. 88 – 49 = _____

5. 76 – 63 = _____

6. $345 – $122 = _____

7. $561 – $377 = _____

8. 505 – 328 = _____

9. 887 – 515 = _____

10. 606 – 444 = _____

11. $492 – $38 = _____

12. $492 – $389 = _____

13. 916 – 588 = _____

14. 903 – 494 = _____

Problem Solving

Use data from the table for problems 15–16.

15. Arturo has $57. He wants to buy a United States singles, a United States blocks, and a Japan album. Does he have enough money?

    Explain. _____

    _____

Stamp Albums

Type	Price
United States singles	$19.95
United States blocks	$26.95
World singles	$39.95
United Kingdom	$22.95
Africa	$19.95
Japan	$19.95

16. Brenda is saving for the Africa stamp album. She has saved $15.96. How much more money does she need? Explain. _____

Spiral Review

Add mentally.

17. 345 + 456 = _____

18. 689 + 223 = _____

19. 406 + 415 = _____

20. 193 + 856 = _____

Name_____

 **2·13 Estimate Differences**

Estimate each difference.

1. 567 – 292 _____   **2.** $1.24 – $0.48 _____

3. $22.98 – $2.57 _____   **4.** $45.47 – $33.21 _____

5. 4,596 – 2,223 _____   **6.** 9,518 – 4,677 _____

7. 45,055 – 5,997 _____   **8.** 672,099 – 54,883 _____

9. 765,902 – 523,111 _____

Subtract. Estimate to check that each answer is reasonable.

10.   568   **11.**   345,600   **12.**   890,906
       – 345   – 221,099   – 591,313

13.   802,586   **14.**   996,517
       – 444,441   –   2,069

Problem Solving

15. Janelle collects dolls from around the world. She has 162 countries on her list. So far she has dolls from 74 different countries. About how many more does she need? _____

16. Melba also collects dolls. She has 158 countries on her list. So far she has 92. About how many more does she need? _____

17. Margit collects miniature flags of nations. She has 54 so far. She has 101 more on her list. How many were on her list to begin with? _____

Spiral Review

Order from greatest to least.

18. 12,356; 12,563; 15,632 _____

19. 902,445; 904,245; 902,454 _____

20. 381,087; 318,078; 387,081 _____

Name_____

3·1 Tell Time

Write the time in two ways. Use A.M. or P.M. when possible.

1.

2.

3.

_____   _____   _____

_____   _____   _____

_____   _____   _____

Tell how much time.

4. 45 minutes = _____

5. 90 minutes = _____

Problem Solving

6. Milton spent an hour practicing his tuba. Then he did his homework for a half hour. Next he took 15 minutes setting the table for dinner. How much time passed?

_____

7. Kendra spent 30 minutes weeding the flower bed. Then she spent a quarter hour preparing the vegetable patch. Next she spent another half hour planting. How much time passed?

_____

Spiral Review

Estimate.

8. 679 + 592 _____

9. 2,670 − 1,992 _____

10. 134,500 + 165,690 _____

11. 599,546 − 172,640 _____

Name_____

3·2 Elapsed Time

How much time has passed?

1. Begin: 7:30 A.M.
   End: 10:00 A.M.

   _____

2. Begin: 5:56 P.M.
   End: 9:00 P.M.

   _____

3. Begin: 11:20 A.M.
   End: 1:06 P.M.

   _____

4. Begin: 12:30 P.M.
   End: 4:15 P.M.

   _____

5. Begin: 9:55 A.M.
   End: 4:45 P.M.

   _____

6. Begin: 8:00 A.M.
   End: 12:00 A.M.

   _____

What time will it be in 30 minutes?

7. **7:45** AM PM

   _____

8. **12:51** AM PM

   _____

9. **4:16** AM PM

   _____

Problem Solving

10. Vanetta and her family are going to visit her grandmother. They start at 10:15 A.M., and the trip takes them 5 hours 20 minutes. When do they arrive? _____

11. Myrna is selling cookies over the telephone. She starts at 11:22 A.M. and finishes after 3 hours 15 minutes. What time did she finish? _____

Spiral Review

Add mentally.

12. 523 + 456 = _____

13. 332 + 626 = _____

14. 219 + 155 = _____

15. 445 + 520 = _____

3·3 Calendar

Use the calendar for problems 1–4.

January						
Sunday	Monday	Tuesday	Wednesday	Thursday	Friday	Saturday
	1	2	3	4	5	6
7	8	9	10	11	12	13
14	15	16	17	18	19	20
21	22	23	24	25	26	27
28	29	30	31			

1. Martin Luther King, Jr.'s, birthday is celebrated on the third Monday in January. What date does it fall on here?

   _____

2. How many days after New Year's Day does it fall?

   _____

   What is another way of saying that?

   _____

3. If your vacation begins on January 14 and ends the following Sunday, how many days long is your vacation?

   _____

4. On what day of the week will the next month start?

   _____

Problem Solving

5. A millennium is 1,000 years. A century is 100 years. A decade is 10 years. How many centuries are there in two millenniums?   _____

6. How many decades are there in a millennium?   _____

Spiral Review

7. $590.06 + $540.02 = _____

8. $458.88 − $359.76 = _____

9. $512.30 + $248.84 = _____

10. $983.18 − $536.33 = _____

11. $658.34 + $296.78 = _____

12. $381.68 − $269.54 = _____

3·4 ▷ Line Plots

1. Make a tally table and a line plot for the following data.

Number of blocks students live from a supermarket:
1, 5, 5, 2, 3, 2, 1, 1, 2, 3, 2, 4, 5, 5, 6, 2, 2, 3, 1, 4

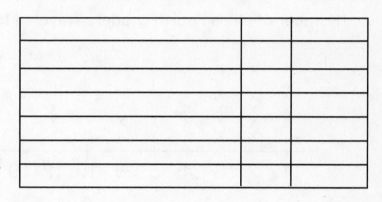

2. What question do you think the survey asked?

_____

_____

_____

_____

3. How many blocks away is the farthest student?

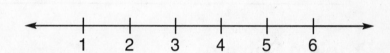

_____

4. How many students took part in this survey? _____

Problem Solving

Use data from the tally chart for problems 5–6.

5. How many students took part in the survey?

_____

6. For how many hours did the most students practice? _____

Hours Students Practice Running During Week

Number of Hours	1	2	3	4	5	6	7
Number of Students	I	IIII	IIIII	IIII I	IIIII	IIII	II

Spiral Review

Round each number to the nearest thousand.

7. 549 _____   **8.** 2,504 _____   **9.** 5,699 _____   **10.** 25,254 _____

3·5 Range, Median, and Mode

Use data from the line plot for problems 1–3.

Number of Miles Fourth Graders Traveled Last Weekend

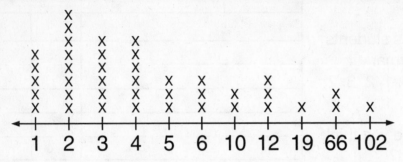

1. What are the median, the mode, and the range for this line plot? _____

2. What does the median tell you about the data? _____

   _____

3. What does the mode tell you about the data? _____

   _____

Problem Solving

Use data from the tally chart for problems 4–7.

4. What are the median, the mode, and the range for this table? _____

5. Which hour had the most visitors entering the park? _____

6. Which hour had the fewest visitors entering the park? _____

Number of Visitors Entering State Park by Hour	
9:00 A.M.	20
10:00 A.M.	60
11:00 A.M.	85
12:00 P.M.	90
1:00 P.M.	90
2:00 P.M.	100
3:00 P.M.	50
4:00 P.M.	40
5:00 P.M.	5

Spiral Review

Use mental math to add or subtract.

7. 65 + 56 = _____   8. 82 + 76 = _____   9. 92 − 45 = _____

Name_____

Problem Solving: Reading for Math
Identify Extra and Missing Information

Solve. Identify extra or missing information in each problem.

1. Carrie and her family travel from Los Angeles to San Francisco mainly by the Pacific Coast Highway. It takes them 10 hours. Nadja and her family travel from Los Angeles to San Francisco mainly by Interstate 5. It takes them 8 hours. How much faster did Nadja's family go?

_____

_____

2. It takes Jeremy's family 4 hours and 45 minutes to drive from Houston to Baton Rouge. If they leave at 5:15, what time will they arrive?

_____

Use data from the table for problems 3–4.

MILEAGE BETWEEN FLORIDA CITIES

	Fort Myers	Jacksonville	Miami	Orlando	Tallahassee	Tampa
Fort Myers		298	152	156	398	129
Jacksonville	298		341	142	163	202
Miami	152	341		229	478	273
Orlando	156	142	229		257	84
Tallahassee	398	163	478	257		275
Tampa	129	202	223	84	275	

3. How long does it take to get from Tallahassee to Miami?

_____

4. Order the other five cities by distance from Orlando, starting with the nearest.

_____

Spiral Review

Round these distances to the nearest ten miles and then the nearest hundred miles.

5. 455 miles _____

6. 176 miles _____

7. 92 miles _____

Name_____

3·7 **Problem Solving: Strategy**
Work Backward

Work backward to solve.

1. María babysits for her cousin after school. It takes her
   25 minutes to walk to her cousin's house, and she arrives
   at her cousin's at 3:20 P.M. What time did she leave school? _____

2. Biff works in his uncle's grocery after school. Martin came
   in and spent $2.50 on soda, $2.99 on cheese, and $1.98
   on cupcakes. He said to Biff, "Now I have $12.00 left in
   my wallet." How much money did he start with? _____

Mixed Strategy Review

3. Calvin rides 4 miles a day for his paper route. How many
   miles does he ride in March? _____

4. Leslie spends $1.35 for breakfast and $4.75 for lunch
   each day from Monday through Friday. How much does
   she spend for her meals during this time? _____

Use data from the table for problems 5–6.

5. What is the attendance range?

   _____

6. On which days did attendance go
   higher than 45,000?

   _____

**Attendance at Houston
Astros Baseball Games**

May 26	36,788
May 27	40,990
May 28	42,713
May 29	46,599
May 30	51,709
May 31	49,380

7. Bridget and Andrea go to the game
   with their father. It takes 1 hour and
   5 minutes to drive there, 20 minutes to park, and 15 minutes to get to their
   seats. If the game starts at 2:00 P.M., what time should they leave the house?

   _____

Spiral Review

Subtract. Check your work by adding.

8. 590,002 – 343,565 = _____   9. 400,600 – 120,993 = _____

Name_____

3·8 Explore Pictographs

Use data from the table for problems 1–5.

Country
I Most Want to Visit

Country	Number of Votes
Brazil	15
England	50
France	45
Japan	30
Mexico	35
Sweden	20

Country I Most Want to Visit

Brazil	✈ ✈
England	✈ ✈ ✈ ✈ ✈
France	
Japan	
Mexico	
Sweden	

Key: Each ✈ stands for 10 votes.

Each ✈ stands for 5 votes.

1. Complete the pictograph. _____

2. Which country is first choice? _____

3. How many more votes did England get than Brazil? How do you know?
   _____

4. How would this pictograph change if you used 1 airplane for 5 votes?
   _____

5. How many voters voted in this survey? _____

Solve.

6. On a pictograph, each 🏠 stands for 5 houses. How
   would you show 45 houses?   _____

Spiral Review

7. It is 6:00 P.M. and James just got off work. If he
   worked 7 hours 30 minutes and took a half hour
   for lunch, what time did he start?   _____

8. Dahlia arrived in Visalia at 3:35 P.M. She had been
   riding since 10:55 A.M. How long was her trip?   _____

Name_____

3·9 Bar Graphs

Use data from the bar graph for problems 1–2.

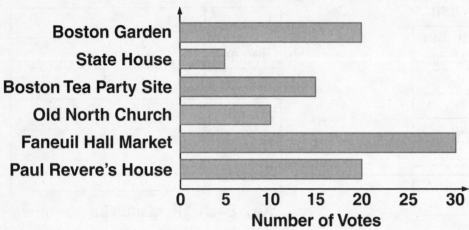

Students' Favorite Places Along the Freedom Trail

1. If each student voted once, how many students were surveyed? _____

2. Which places got the same number of votes? How many?

_____

Problem Solving

Use data from the table for problems 3–4.

3. If you wanted to make a bar graph using the data in the table, what is a good scale to use? Why?

_____

_____

_____

4. What is the range of the data? _____

Number of Students Who Have Visited Each City

City	Boys	Girls
Dallas	4	6
Houston	6	4
Las Vegas	10	8
Los Angeles	8	10
Phoenix	4	4
San Francisco	6	6

Spiral Review

5. $5.92 − $4.55 = _____

6. $6.62 + $3.04 + $2.12 = _____

7. $90.05 − $59.99 = _____

8. $19.95 − $18.99 = _____

3·10 Coordinate Graphing

Give the ordered pairs for each of these places on the grid.

1. Admission Booth

_____

2. Mighty Mammoths

_____

3. Flora Fossils

_____

4. Flying Dragon

_____

5. Sea Monsters

_____

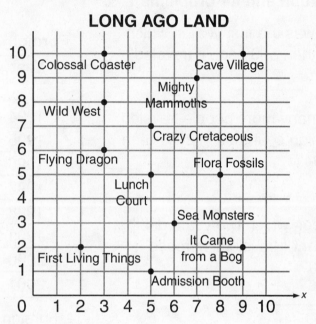

LONG AGO LAND

Name the place at each location.

6. (3, 10) _____   **7.** (5, 5) _____

8. (9, 10) _____   **9.** (5, 7) _____

Problem Solving

10. If each grid line stands for 100 feet, how far is it from
the Admissions Booth to Crazy Cretaceous?   _____

11. How far is it from the Colossal Coaster to the Cave Village?   _____

Spiral Review

Tell how much time has passed.

12. Begin: 9:15 a.m.
End: 4:45 p.m.

13. Begin: 12:05 a.m.
End: 11:05 p.m.

14. Begin: 11:05 a.m.
End: 12:23 p.m.

_____   _____   _____

Name_____

3·11 **Explore Line Graphs**

Use data from the table to complete the line graph and for problems 1–5.

1. What was the first year in which more than 500 people traveled?

   _____

2. How many more people traveled to foreign countries in 2001 than in 1994?

   _____

3. Between which two years did the number of travelers increase by 200?

   _____

4. Between which two years did the number of travelers not increase by as many as between the previous two years?

   _____

5. Based on the graph, about how many townspeople do you expect will travel to foreign countries in 2002?

   _____

   _____

Foreign Travel by Town Residents

Year	Number Who Traveled
1994	125
1995	167
1996	202
1997	265
1998	370
1999	515
2000	715
2001	975

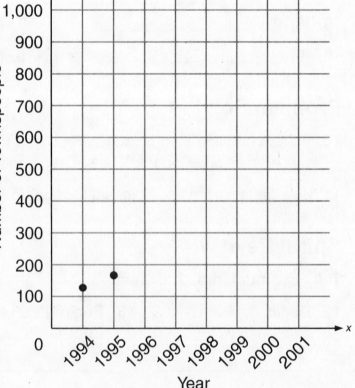

Foreign Travel by Town Residents

Spiral Review

Round to the nearest 10 cents.

6. $5.64 _____

7. $6.92 _____

 **4·1** **The Meaning of Multiplication**

Find each product.

1. 9
    × 1

2. 7
    × 5

3. 7
    × 4

4. 5
    × 6

5. 8
    × 4

6. 6
    × 7

7. 5
    × 8

8. 4
    × 3

9. 8
    × 2

10. $2 \times 5 =$ _____

11. $3 \times 7 =$ _____

12. $4 \times 6 =$ _____

13. $9 \times 2 =$ _____

14. $4 \times 4 =$ _____

15. $1 \times 6 =$ _____

Problem Solving

16. Karen and Paul both collect dinosaurs. Karen has 5 shelves with 6 dinosaurs on a shelf. Paul has 4 shelves with 8 dinosaurs on a shelf. Who has more dinosaurs? How many more?

_____

17. Vern is arranging a display table with doorknobs. If he places 6 doorknobs in each of 4 rows, how many will he have?

_____

Spiral Review

Give the value of each underlined digit.

18. 5,536 _____

19. 3,064 _____

20. 76,942 _____

21. 45,336,990 _____

22. 4,932,115 _____

 **4·2** **Properties of Multiplication**

Multiply. Then use the Commutative Property to write a different multiplication sentence.

1. $6 \times 0 =$ _____   **2.** $5 \times 7 =$ _____

3. $8 \times 2 =$ _____   **4.** $5 \times 3 =$ _____

5. $3 \times 3 =$ _____   **6.** $1 \times 4 =$ _____

7. $6 \times 2 =$ _____   **8.** $0 \times 8 =$ _____

9. $2 \times 9 =$ _____   **10.** $1 \times 8 =$ _____

11. $8 \times 4 =$ _____   **12.** $2 \times 4 =$ _____

13. $9 \times 3 =$ _____   **14.** $4 \times 9 =$ _____

15. $9 \times 0 =$ _____

Problem Solving

16. Randa is stringing beads in her sister's hair. If she puts 5 beads on each row and there are 6 rows, how many beads does she need?   _____

17. Barrie is looking at the collection of knights' helmets in the museum. If there are 3 rows with 6 helmets each, how many helmets are there?   _____

18. Ping likes to play with his brother's toy soldiers. He takes out 35 toy soldiers to play with. This leaves 46 soldiers in the box. How many soldiers were in the box to start with?   _____

Spiral Review

19. $545,602 + 413 =$ _____   **20.** $2,195 + 3,337 =$ _____

21. $45,389 + 40,001 =$ _____   **22.** $675 + 92 =$ _____

Name_____

Multiply by 2, 3, 4, and 6

Write a multiplication sentence to solve.

1.

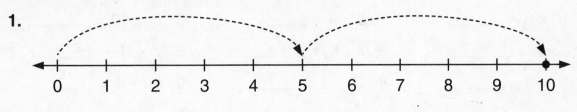

0   1   2   3   4   5   6   7   8   9   10

_____

Multiply.

2. $6 \times 5 =$ _____   **3.** $3 \times 5 =$ _____   **4.** $4 \times 2 =$ _____

5. $6 \times 9 =$ _____   **6.** $3 \times 2 =$ _____   **7.** $4 \times 8 =$ _____

8. $4 \times 3 =$ _____   **9.** $6 \times 4 =$ _____   **10.** $4 \times 5 =$ _____

11. 8   **12.** 8   **13.** 7   **14.** 0
    $\times 6$ $\times 3$ $\times 3$ $\times 4$

Problem Solving

> ● SPECIAL: Old Stamps $2 a Packet. ●

15. Old stamps come in a packet of 4. If Zara buys 6 packets, how many stamps will she have? How much change will she get from $20?   _____

16. Musa's father owns the stamp store. He has 52,416 stamps in stock. How many more stamps does he need to have 60,000 stamps?   _____

Spiral Review

17. $56,997 - 42,130 =$ _____   **18.** $479,089 - 38,668 =$ _____

19. $3,870,221 - 433,675 =$ _____   **20.** $24,569 - 194 =$ _____

Name_____

 **4·4** **Multiply by 5 and 10**

Multiply.

1. $10 \times 9 =$ _____   2. $5 \times 9 =$ _____   3. $5 \times 10 =$ _____

4. $7 \times 8 =$ _____   5. $7 \times 6 =$ _____   6. $7 \times 10 =$ _____

7. $10 \times 3 =$ _____   8. $10 \times 2 =$ _____   9. $10 \times 6 =$ _____

10. $\begin{array}{r} 9 \\ \times\, 10 \\ \hline \end{array}$ 11. $\begin{array}{r} 9 \\ \times\, 5 \\ \hline \end{array}$ 12. $\begin{array}{r} 8 \\ \times\, 5 \\ \hline \end{array}$

Tell whether the number is a multiple of 2, 5, or 10.

13. 16 _____   14. 40 _____   15. 34 _____

16. 50 _____   17. 35 _____   18. 65 _____

Problem Solving

19. Tanya's display case has 7 rows of 5 boxes each. Arthur's display case has 4 rows of 10 boxes each. Whose display case has more boxes? How do you know?

_____

20. Roberta brought her arrowhead collection to school. There are 5 rows of 6 arrowheads each in her display case. Luis brought his miniature car collection. He has 7 rows with 4 cars each. Does Roberta have more arrowheads than Luis has cars? How do you know?

_____

Spiral Review

21. $889 - 90 =$ _____   22. $6{,}718 - 503 =$ _____

23. $3{,}410 - 2{,}675 =$ _____   24. $5{,}677 - 4{,}199 =$ _____

Name_____

 **4·5** **Multiply by 7, 8, and 9**

Multiply.

1. $7 \times 3 =$ _____   2. $7 \times 9 =$ _____   3. $8 \times 7 =$ _____

4. $8 \times 0 =$ _____   5. $5 \times 8 =$ _____   6. $2 \times 9 =$ _____

7. 10   8.   0   9.   9   10.   8
    ×9   ×7   ×6   ×4

11.   5   12.   9   13.   7   14.   8
     ×7   ×8   ×6   ×1

Problem Solving

15. Francis has begun to collect old records. His aunt Flo gave him some money, and he bought 8 records in each of 4 different stores. Then she gave him 7 of her records. How many records does Francis have?   _____

16. Minerva's CD rack has 8 shelves. Each shelf can hold 9 CDs. How many CDs can the rack hold in all?   _____

17. Akiko has the same CD rack as Minerva. She has filled 7 shelves. How many CDs does she have?   _____

18. Kenan also has the same CD rack. He has filled 3 shelves, and there are 5 CDs on another shelf. How many CDs does he have?   _____

Spiral Review

Write in expanded form:

19. 47,192 _____

20. 22,018 _____

21. 34,106 _____

22. 56,778 _____

4·6 **Problem Solving: Reading for Math**

Choose an Operation

Solve. Tell how you chose the operation.

1. Emma has 520 pencils in her collection. Her friend Eve has 264. How many more pencils does Emma have than Eve? What is the total number of pencils that they have between them?

   _____

   _____

2. The gift shop at the aquarium sells gift packages containing 8 pencils each. If Ms. Martin wants to buy a package for each student in her class of 24, how many pencils will she have bought in all?

   _____

   _____

Use data from the table for problems 3–5.

3. If Dorothy buys 36 pocket folders, how many will be left?

   _____

4. If pencils come 10 to a package, how many packages are there in stock?

   _____

5. How many pencils and pens are there in all?

   _____

   _____

In Stock at The Write Time Stationery Store

Pencils	800
Pens	650
1-pen / 1-pencil sets	156
Boxed stationery	300
Plain envelopes (boxes)	124
Pocket folders	420

Spiral Review

6. $678 - 344 =$ _____

7. $3,657 - 569 =$ _____

8. $45,112 - 5,620 =$ _____

9. $555,111 - 29,399 =$ _____

Name_____

 **4·7** **Multiplication Table and Patterns**

Multiply.

1. $4 \times 12 =$ _____

2. $8 \times 3 =$ _____

3. $7 \times 6 =$ _____

4. $8 \times 6 =$ _____

5. $11 \times 12 =$ _____

6. $6 \times 10 =$ _____

7. $5 \times 9 =$ _____

8. $8 \times 4 =$ _____

0	0	0	0	0	0	0	0	0	0	0	0	0
1	0	1	2	3	4	5	6	7	8	9	10	11	12
2	0	2	4	6	8	10	12	14	16	18	20	22	24
3	0	3	6	9	12	15	18	21	24	27	30	33	36
4	0	4	8	12	16	20	24	28	32	36	40	44	48
5	0	5	10	15	20	25	30	35	40	45	50	55	60
6	0	6	12	18	24	30	36	42	48	54	60	66	72
7	0	7	14	21	28	35	42	49	56	63	70	77	84
8	0	8	16	24	32	40	48	56	64	72	80	88	96
9	0	9	18	27	36	45	54	63	72	81	90	99	108
10	0	10	20	30	40	50	60	70	80	90	100	110	120
11	0	11	22	33	44	55	66	77	88	99	110	121	132
12	0	12	24	36	48	60	72	84	96	108	120	132	144
	0	1	2	3	4	5	6	7	8	9	10	11	12

9.   5
   $\times 11$

10.   12
   $\times 8$

11.   7
   $\times 9$

12.   9
   $\times 8$

Problem Solving

13. What pattern do you see in the 3 and 9 columns? _____

_____

14. What pattern do you notice about the square numbers (the numbers in dots)?

_____

Spiral Review

15. $145 + 136 + 120 =$ _____   16. $5,662 - 293 =$ _____

4·8 Multiply Three Numbers

Multiply.

1. $(7 \times 2) \times 4 =$ _____

2. $5 \times (9 \times 0) =$ _____

3. $8 \times (2 \times 5) =$ _____

4. $(2 \times 9) \times 2 =$ _____

5. $(3 \times 2) \times 3 =$ _____

6. $(7 \times 1) \times 4 =$ _____

7. $(8 \times 2) \times 4 =$ _____

8. $(1 \times 5) \times 7 =$ _____

9. $3 \times (3 \times 0) =$ _____

10. $(4 \times 5) \times 4 =$ _____

Complete the multiplication sentence.

11. $(5 \times 6) \times 3 =$ _____

12. $7 \times (1 \times 8) =$ _____

13. $5 \times 8 =$ _____ $\times 5$

14. $9 \times 6 =$ _____ $\times 9$

Problem Solving

Use the data from the table for problems 15–17.

15. How many more cars does Zak have than Tom?

_____

16. Which two collectors have about the same number of cars?

_____

17. How many more cars does Ann need to have the same number as Don?

_____

Model Car Collectors

Name	Number of Cars
Ann	31
Don	52
Pat	78
Tom	61
Zak	81

Spiral Review

18. Order least to greatest: 451, 445, 441, 454, 545

_____

19. Order greatest to least: 889, 989, 988, 888, 998

_____

4·9 Relate Multiplication and Division Facts

Divide.

1. 15 ÷ 5 = _____

2. 27 ÷ 9 = _____

3. 63 ÷ 7 = _____

4. 48 ÷ 8 = _____

5. 16 ÷ 4 = _____

6. 25 ÷ 5 = _____

7. 9 ÷ 1 = _____

8. 33 ÷ 3 = _____

9. 90 ÷ 10 = _____

10. 5)‾50‾

11. 7)‾49‾

12. 2)‾24‾

13. 6)‾54‾

14. 8)‾32‾

15. 7)‾56‾

16. 9)‾36‾

17. 8)‾80‾

18. 9)‾108‾

19. 2)‾18‾

20. 6)‾36‾

21. 7)‾77‾

22. 4)‾32‾

23. 4)‾24‾

24. 8)‾64‾

Problem Solving

25. Marilyn has a collection of 48 snow globes. If she divides them evenly among 6 shelves, how many globes will there be on each shelf? _____

26. If Marilyn gets 2 more shelves, how many snow globes will fit evenly on each shelf? _____

Spiral Review

27. 3 × 4 × 5 = _____

28. 2 × 0 × 8 = _____

29. 3 × 2 × 2 = _____

30. 4 × 4 × 4 = _____

31. 1 × 5 × 5 = _____

32. 7 × 6 × 1 = _____

 **4·10** **Problem Solving: Strategy**

Act It Out

Solve.

1. The school is having a bake sale to raise money for new gym equipment. Yolanda is planning to bring 48 cupcakes. She is buying bakery boxes to carry them in. There are boxes that will hold 9 cupcakes each, and there are boxes that will hold 12 cupcakes each. If she wants to fill all the boxes she buys, which size boxes should she buy? How many? _____

2. Manolo is bringing donut holes. He figures he will sell them in groups of 3. If he wants to buy enough for 12 customers, should he get a bag of 20 or 40? How many will he have left? _____

Mixed Strategy Review

Use data from the table for problems 3–5.

3. How many more collectors are there in the country with the most collectors than in the country with the fewest collectors?

    _____

Model Car Collectors

Country	Number of Collectors
Australia	10,041
Canada	11,044
France	35,900
United Kingdom	24,676
United States	82,399

4. Which country has about 3 times as many collectors as Canada? _____

5. Which two countries have the smallest difference in their numbers of collectors? What is the difference? _____

Spiral Review

6. 675,902 − 568,092 = _____   7. 1,003,094 − 717,003 = _____

Name_____

 **4·11** **Divide by 2 Through 12**

Divide.

1. 14 ÷ 7 = _____   **2.** 81 ÷ 9 = _____   **3.** 30 ÷ 10 = _____

4. 96 ÷ 8 = _____   **5.** 72 ÷ 12 = _____   **6.** 21 ÷ 7 = _____

7. 4)‾28‾   **8.** 11)‾121‾   **9.** 6)‾54‾   **10.** 7)‾77‾

Problem Solving

Use data from the line plot for problems 11–12.

The Toy Horse Collectors' Club took a survey of the ages of its members. Here are the results:

Ages of Club Members

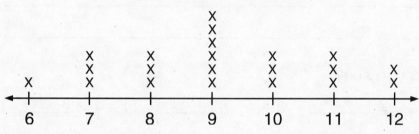

11. What is the median? _____   **12.** What is the mode? _____

Solve.

13. Johanna's grandfather has a collection of 72 cuckoo clocks. He plans to give an equal number to each of his children and grandchildren. If he has 4 children and 8 grandchildren, how many clocks will each get? _____

14. Johanna's older cousin Margo does not care for cuckoo clocks and plans to divide her share equally among her three children. How many clocks will each of them get? _____

Spiral Review

15. 3 × 7 × 3 = _____   **16.** 8 × 5 × 2 = _____

Name_____

 **4·12** **Fact Families**

Complete each fact family.

1. $8 \times 9 = a$ _____
   $9 \times b = 72$ _____
   $72 \div 8 = c$ _____
   $d \div 9 = 8$ _____

2. $8 \times 8 = e$ _____
   $64 \div f = f$ _____

3. $4 \times 11 = g$ _____
   $11 \times h = 44$ _____
   $44 \div 4 = j$ _____
   $k \div 11 = 4$ _____

Find each missing factor.

4. $7 \times m = 28$ _____
   $28 \div 4 = n$ _____

5. $5 \times p = 40$ _____
   $40 \div 8 = q$ _____

6. $10 \times r = 30$ _____
   $30 \div 3 = s$ _____

Write a multiplication and division fact family for each group of numbers.

7. 9, 11, 99 _____

8. 8, 7, 56 _____

Divide. What patterns do you see?

9. $3 \div 1 =$ _____ $6 \div 1 =$ _____ $8 \div 1 =$ _____ $11 \div 1 =$ _____

_____

10. $2 \div 2 =$ _____ $4 \div 4 =$ _____ $9 \div 9 =$ _____ $12 \div 12 =$ _____

_____

Problem Solving

11. Warren raises prize rabbits. If he has 7 females and each one gives birth to 4 baby bunnies, how many baby bunnies will there be? Write a number sentence to show this.   _____

12. At the fair, Marisol buys a doll for $14.50 and a pretzel for $1.25. How much change will she receive from a $20 bill?   _____

Spiral Review

13. $5 \times 5 \times 4 =$ _____   14. $9 \times 2 \times 3 =$ _____

Name_____

5·1 ► Patterns of Multiplication

Complete.

1. $6 \times 7 = a$ _____

   $6 \times b = 420$ _____

   $c \times 700 = 4,200$ _____

   $6 \times 7,000 = d$ _____

2. $8 \times 5 = a$ _____

   $8 \times b = 400$ _____

   $c \times 500 = 4,000$ _____

   $8 \times 5,000 = d$ _____

3. $9 \times 4 = a$ _____

   $9 \times b = 360$ _____

   $c \times 400 = 3,600$ _____

   $9 \times 4,000 = d$ _____

4. $7 \times 3 = a$ _____

   $7 \times b = 210$ _____

   $c \times 300 = 2,100$ _____

   $7 \times 3,000 = d$ _____

Multiply. Use mental math.

5.
$$\begin{array}{r} 70 \\ \times\ 9 \\ \hline \end{array}$$

6.
$$\begin{array}{r} 20 \\ \times\ 7 \\ \hline \end{array}$$

7.
$$\begin{array}{r} 40 \\ \times\ 8 \\ \hline \end{array}$$

8.
$$\begin{array}{r} 500 \\ \times\ \ 7 \\ \hline \end{array}$$

9. $3 \times 800 =$ _____

10. $700 \times 7 =$ _____

11. $600 \times 7 =$ _____

Problem Solving

12. The distance by car from New York City to Grand Canyon National Park, Arizona, is about 2,400 miles. How far is a round trip?

    _____

13. The distance by car from New York City to Crater Lake National Park, Oregon, is about 2,900 miles. How far is a round trip? How much farther is a round trip to Crater Lake than to the Grand Canyon?

    _____

Spiral Review

14. $8 \times 6 =$ _____

15. $12 \times 12 =$ _____

16. $99 \div 11 =$ _____

Name_____

Find each product.

1. 53
    × 6

2. 47
    × 5

3. 29
    × 4

4. 54
    × 8

5. 67
    × 4

6. 22
    × 8

7. 63
    × 8

8. 37
    × 6

9. 19
    × 9

10. 25
    × 7

11. 14
    × 7

12. 58
    × 6

13. 36
    × 5

14. 72
    × 8

15. 26
    × 9

16. $8 \times 56 =$ _____

17. $7 \times 22 =$ _____

18. $7 \times 24 =$ _____

19. $6 \times 52 =$ _____

20. $9 \times 87 =$ _____

21. $4 \times 54 =$ _____

22. $3 \times 37 =$ _____

23. $4 \times 88 =$ _____

24. $5 \times 73 =$ _____

25. $4 \times 66 =$ _____

Spiral Review

26. $8 \times 2 \times 3 =$ _____

27. $81 \div 9 =$ _____

28. $5 \times 0 \times 7 =$ _____

29. $5 \times 1 \times 7 =$ _____

30. $72 \div 8 =$ _____

31. $6 \times 2 \times 4 =$ _____

32. $4 \times 9 \times 1 =$ _____

33. $3 \times 1 \times 8 =$ _____

Name_____

5·3 Multiply by 1-Digit Numbers

Multiply.

1. 53
    × 6

2. 47
    × 5

3. 29
    × 4

4. 54
    × 8

5. 67
    × 4

6. 22
    × 8

7. 8 × 56 = _____

8. 7 × 22 = _____

9. 7 × 24 = _____

10. 6 × 52 = _____

11. 9 × 87 = _____

12. 4 × 54 = _____

Problem Solving

13. There are 48 students in the fourth grade. Each student was asked to collect three autumn leaves. How many leaves did the fourth grade students collect in all? _____

14. Mario and Carlo decide to pool their money and buy their grandmother a yucca plant. The plant costs $12.98 and a dish to put it on costs $4.98. If they pay for the plant and dish with a $20 bill, how much change do they get? _____

Use data from the table for problems 15–16.

15. How many Jaguar Bars are there in all?

    _____

16. Are there more Cougar Bars than there are Jaguar Bars? If so, how many more?

    _____

Big Cat Granola Bars Sold

Kind of Bar	Number of Boxes	Number in Each Box
Bobcat	19	8
Cougar	25	8
Jaguar	57	6
Tiger	38	6

Spiral Review

17. 15,627 + 2,955 = _____

18. 977,606 + 546,998 = _____

19. 581,782 − 443,187 = _____

20. 149,000 − 76,599 = _____

Name_____

5·4 Estimate Products

Estimate each product.

1. 6×57 _____

2. $4 \times \$88$ _____

3. 7×95 _____

4. 19×7 _____

5. $3 \times \$82$ _____

6. 5×79 _____

7. 77×8 _____

8. 29×6 _____

9. 55×8 _____

10. 57×8 _____

11. 8×23 _____

12. 4×229 _____

13. $9 \times \$320$ _____

14. $5 \times \$209$ _____

15. $7 \times 1,455$ _____

Problem Solving

16. At the snack bar in the park ranger station, juice is $.90 a box. If Selma, her parents, and her three brothers each have a juice box, estimate how much money they need.

_____

17. Selma and her family stop at a fast-food restaurant just outside the park. Lunches for all of them cost $3.77 each. Selma's father takes out a $20 bill to pay for the food. Will he have to go back into his wallet for more money?

_____

18. Later, the family stops for ice cream. Cones for all of them cost $1.59 each. If Selma's father pays with a $10 bill, will he have enough?

_____

Spiral Review

19. $7 \times 8 \times 2 =$ _____

20. $3 \times 4 \times 6 =$ _____

21. $18 \div 2 =$ _____

22. $28 \div 4 =$ _____

23. $6 \times 7 \times 2 =$ _____

24. $48 \div 6 =$ _____

Name_____

Problem Solving: Reading for Math

Use an Overestimate or Underestimate

Solve. Tell how you estimated.

1. A boat trip to the wetlands preserve costs $12. Leila's family includes 8 people. They brought $120 to pay for the boat ride. Do they have enough money?

   _____

2. The boat holds 44 people. Are 5 trips enough if 200 people want to ride the boat today?

   _____

3. On the boat, books of discount coupons for the wetlands preserve are given out. If the coupons come in packs of 20, are 12 packs enough for all 200 people riding the boat today?

   _____

4. Millicent has $10 to spend at the wetlands preserve. If she has a coupon for $1 off lunch and lunch costs $5.89, will she have enough money left over to buy a stuffed toy for $6.98?

   _____

5. Water taxis take people through the preserve. If a water taxi can hold 5 people besides the driver, will 11 water taxis be enough to fit all 44 people who arrive on one boat?

   _____

6. The big screen film room holds 54 people. If there are 4 showings, will there be enough room for all 200 visitors today?

   _____

Spiral Review

Round.

7. 546 to tens _____

8. 7,925 to hundreds _____

9. 3,441 to thousands _____

10. 12,756 to ten thousands _____

Name_____

5·6 Multiply Greater Numbers

Multiply. Check for reasonableness.

1. 806 × 8	2. 716 × 4	3. 725 × 7	4. $5.29 × 8	5. 4,559 × 8

6. $6,772 × 9	7. $50.94 × 6	8. 5,554 × 7	9. 6,936 × 3	10. $7,114 × 5

11. $6 \times 56,203 =$ _____

12. $5 \times 4,388 =$ _____

13. $8 \times \$190.53 =$ _____

14. $9 \times \$1,225 =$ _____

15. $6 \times 23,099$ _____

16. $5 \times \$245.67 =$ _____

17. $8 \times 28,667 =$ _____

18. $6 \times 21,339 =$ _____

19. $4 \times \$267.89 =$ _____

20. $8 \times 33,848 =$ _____

Problem Solving

21. Rosita's family is going camping for 5 days. It costs them $46 a day for food. How much will their food cost for all 5 days? _____

22. The campsite costs $27.50 a night. How much will it cost the family for 5 nights? _____

Spiral Review

23. $3,384 + 567 + 290 =$ _____

24. $8,565 - 2,276 =$ _____

25. $53 \times 9 =$ _____

26. $78 \times 7 =$ _____

27. $88 \times 8 =$ _____

28. $47 \times 6 =$ _____

29. $9,436 - 4,853 =$ _____

30. $2,983 + 421 + 306 =$ _____

5·7 Problem Solving: Strategy
Find a Pattern

Find a pattern to solve. Describe the pattern.

1. Myron sets a goal for himself to spot 3 birds on the first day, 6 on the second day, 9 on the third day, and so forth. Describe the pattern. How many birds will he spot on the sixth day?

   _____

2. Bao and Frank are learning to mountain-climb. They climb 70 feet the first day, 85 feet the second day, 100 the third day, 115 the fourth day, and 130 the fifth day. Describe the pattern. How many feet will they climb on the tenth day?

   _____

   _____

3. A forest fire burned 100 acres the first day, 300 the second, 900 the third, and 2,700 the fourth. Describe the pattern. If the fire keeps burning at the same rate, how many acres will it burn on the sixth day?

   _____

   _____

Mixed Strategy Review

Use data from the table for problems 4–5.

4. Of which animal were there twice as many as the number of burrowing owls?

   _____

5. Rangers have also spotted 40 pocket gophers in the Reserve. This was half the number of which other animal?

   _____

Number of Animals Spotted in the Saguaro Desert Preserve	
Burrowing owls	50
Chuckwallas	80
Geckos	120
Gila monsters	100

Spiral Review

Estimate the products.

6. 74×9 _____

7. 135×8 _____

8. 76×5 _____

Name_____

Functions and Graphs

Complete the table. Then write an equation.

1. Doron can hike twice as fast as his brother, Ari.

A	1	2	3	4	5
D	2	4			

_____

Complete the table. Then graph the function.

2. For every inch of rain that falls in the desert, the Barlow River rises by 8 inches.

$b = 8r$

r	1	2	3	4	5
b	8	16			

Problem Solving

3. The Museum of Desert Life is having a special fund-raising dinner. Tickets are $20 a person. How much will it cost Carla, her parents, her brother and sister, and two aunts to go? Write an equation to solve.

_____

4. The food, entertainment, and labor cost the museum $9.79 per person. How much will the museum make from Carla's family? Write an equation to solve.

_____

Spiral Review

5. $345.65 + $25.98 = _____

6. $356.98 − $22.96 = _____

7. 5 × $561.29 = _____

8. 6 × $603.39 = _____

9. $486.32 + $56.23 = _____

10. $524.88 − $46.99 = _____

Name_____

 **6·1** **Patterns of Multiplication**

Find each missing number.

1. $7 \times 5 = a$ _____
   $70 \times b = 350$ _____
   $70 \times 50 = c$ _____
   $70 \times 500 = d$ _____

2. $e \times 6 = 54$ _____
   $90 \times 6 = f$ _____
   $g \times 60 = 5,400$ _____
   $90 \times 600 = h$ _____

3. $4 \times 8 = j$ _____
   $k \times 8 = 320$ _____
   $40 \times m = 3,200$ _____
   $400 \times 80 = n$ _____

Multiply. Use mental math.

4. $30 \times 800 =$ _____

5. $90 \times 6,000 =$ _____

6. $5,000 \times 70 =$ _____

7. $500 \times 70 =$ _____

8. $50 \times 80 =$ _____

9. $40 \times 9,000 =$ _____

10. $30 \times 1,000 =$ _____

11. $700 \times 70 =$ _____

12. $600 \times 70 =$ _____

13. $50 \times 4,000 =$ _____

14. $80 \times 70 =$ _____

15. $3,000 \times 60 =$ _____

16. $9,000 \times 30 =$ _____

17. $30 \times 90 =$ _____

18. $10 \times 500 =$ _____

19. $600 \times 60 =$ _____

Problem Solving

20. The state of Rhode Island has about 40 miles of coastline along the Atlantic Ocean. If the entire Atlantic coast of all the states is about 50 times that of Rhode Island, how long is it? _____

21. The United States' Pacific coastline is about 7,600 miles. How much longer is it than the Atlantic coastline? _____

Spiral Review

Compare. Write >, <, or =.

22. 7,336 _____ 73,336

23. 56,989 _____ 56,898

24. 4,643 _____ 4,463

25. 50,029 _____ 5,029

6·2 Explore Multiplying by 2-Digit Numbers

Multiply.

1. $38 \times 41 =$ _____

2. $42 \times 27 =$ _____

3. $33 \times 38 =$ _____

4. $29 \times 21 =$ _____

5. $36 \times 25 =$ _____

6. $20 \times 37 =$ _____

7. $19 \times 39 =$ _____

8. $22 \times 36 =$ _____

9. $14 \times 18 =$ _____

10. $34 \times 23 =$ _____

11. $36 \times 17 =$ _____

12. $12 \times 48 =$ _____

13. $43 \times 20 =$ _____

14. $46 \times 22 =$ _____

15. $35 \times 35 =$ _____

16. $24 \times 32 =$ _____

17. $24 \times 18 =$ _____

18. $17 \times 41 =$ _____

19. $33 \times 15 =$ _____

20. $39 \times 26 =$ _____

21. $40 \times 29 =$ _____

22. $37 \times 36 =$ _____

23. $22 \times 45 =$ _____

24. $13 \times 38 =$ _____

Spiral Review

25. $80 \times 5 =$ _____

$800 \times 50 =$ _____

$8,000 \times 50 =$ _____

$80 \times 50 =$ _____

$8 \times 50 =$ _____

26. $30 \times 4 =$ _____

$300 \times 40 =$ _____

$3000 \times 40 =$ _____

$30 \times 40 =$ _____

$3 \times 40 =$ _____

6·3 Multiply by Multiples of 10

Multiply.

1. 53
    × 10

2. 47
    × 50

3. 29
    × 40

4. 54
    × 80

5. 67
    × 40

6. 22
    × 50

7. 296
    × 40

8. 544
    × 80

9. 6,710
    ×   40

10. 2,254
    ×   80

11. 6,056
    ×   60

12. 2,527
    ×   70

13. $48 \times 50 =$ _____

14. $70 \times 22 =$ _____

15. $90 \times 24 =$ _____

16. $56 \times 50 =$ _____

17. $90 \times 872 =$ _____

18. $448 \times 50 =$ _____

19. $8,134 \times 50 =$ _____

20. $20 \times 7,172 =$ _____

Problem Solving

21. A ferry boat can carry 286 passengers per trip. How many passengers can the ferry carry in a weekend if it makes 20 trips?   _____

22. A barge can carry 385 tons of coal. How much coal can it carry in a year if it makes 50 trips?   _____

Spiral Review

23. $313,560 + 345,999 =$ _____

24. $986,905 - 885,040 =$ _____

25. $906 \times 5 =$ _____

26. $8 \times 5 \times 8 =$ _____

6·4   **Problem Solving: Reading for Math**

Solve Multistep Problems

Solve. Tell what inference you made.

1. During the summer beach season, Brigid and Mitch were lifeguards. Brigid worked 4 hours a day, five days a week, and Mitch worked twice as many hours a week as Brigid. The season lasted 12 weeks. How many hours did each of them work during the season?

_____

_____

Use data from the graph for problems 2–3. Where you have to infer something to solve the problem, tell what it is.

Cindy's parents run a boarding house on the beach. This graph shows the number of "room rental days" from last summer. A "room rental day" is any day that a room is occupied. It is different from the number of guests.

2. If the boarding house has 10 rooms, in which months were all the rooms occupied?

_____

_____

_____

3. If the rent for a room is $50 a night, estimate how much money Cindy's parents take in during a month when the boarding house is full.

_____

_____

_____

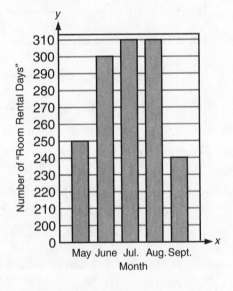

Spiral Review

4. $45 \times 40 =$ _____

5. $42 \times 43 =$ _____

6. $48 \times 37 =$ _____

Name_____

6·5 Multiply by 2-Digit Numbers

Find each product.

1. 92
    × 47

2. 29
    × 86

3. $0.77
    × 40

4. 58
    × 61

5. 52 × $0.99 = _____

6. 84 × 27 = _____

7. 23 × 59 = _____

8. 26 × 48 × 2 = _____

9. 52 × 36 × 3 = _____

Problem Solving

10. A ship leaving a river port can carry 25 tons of cargo. If the ship carries a full cargo and makes 46 trips in a year, how much cargo can it carry from port to port in a year? _____

Use data from the table for problems 11–13.

11. List the rivers in order from greatest to shortest length.

    _____

    _____

    _____

Length in Miles of Some U.S. Rivers	
Arkansas	1,459
Colorado	1,450
Mississippi	2,340
Missouri	2,315
Ohio	981
Rio Grande	1,900
Tennessee	652

12. Which two pairs of rivers have similar lengths?

    _____

    _____

13. Which river is about twice as long as the Ohio? _____

Spiral Review

14. 9 × 7 × 6 = _____

15. 4 × 8 × 8 = _____

16. 9 × 8 × 2 = _____

Name_____

Estimate each product. Tell how you rounded.

1. 77 × 94 = _____

2. 405 × 67 = _____

3. 333 × 250 = _____

4. 17 × 59 = _____

5. 56 × 548 = _____

6. 47 × $5,698 = _____

7. 898 × 15 = _____

8. 992 × $135 = _____

9. 561 × 22 = _____

10. 925 × 443 = _____

11. 36 × $583 = _____

12. 135 × 496 = _____

13. 579 × 23 = _____

14. 796 × 33 = _____

15. 369 × 109 = _____

Problem Solving

16. The Alpha Company makes rubber rafts that sell for $29.95 each. If the Rapture River Rafting Regatta buys 42 rafts, about how much money do they spend? _____

17. The Rafting Regatta charges $20 a trip and sold 2,019 trips last month. About how much money did it collect? _____

Spiral Review

18. 450 + 379 + 356 = _____

19. 90,825 − 4,776 = _____

20. 45 × $0.93 = _____

Name_____

Multiply. Check that each answer is reasonable.

1. $5.37 × 42	**2.** 472 × 59	**3.** $8.29 × 64
4. 754 × 66	**5.** 679 × 55	**6.** 3,322 × 95
7. 4,296 × 83	**8.** $6,544 × 88	**9.** 56,117 × 44

10. $73 \times 47,096 =$ _____

11. $32 \times \$159.95 =$ _____

Problem Solving

Use data from the sign for problems 12–13.

12. Mary Lou has a monthly student pool membership. If she pays the monthly locker fee and brings 4 guests during the month, but they don't use a separate locker, how much does she spend at the pool this month? _____

POOL FEES

Monthly membership—adult	$35.00
Monthly membership—student	$20.50
Monthly membership—65 and over	$25.00
Guests—each one, daily	$2.25
Locker—monthly	$15.75
Locker—daily	$1.75

13. If Mary Lou's classmate Roberta plans to go to the pool with Mary Lou 3 times a week for the whole month, will it be cheaper for her to buy a monthly membership? Explain your answer.

_____

_____

Spiral Review

14. Carlos swims laps in the pool every weekday. If he swims 48 laps each day, about how many laps does he swim in a month? Tell what you need to infer, and tell how you rounded.

_____

_____

15. If each lap is 96 feet, about how many feet does Carlos swim in a month?

_____

Name_____

Problem Solving: Strategy

Make a Graph

Use data in the table to solve problems 1–5.

**Average Difference
Between High and Low Tides**

Boston, MA	10 feet 4 inches
Charleston, SC	5 feet 10 inches
Fort Pulaski, GA	7 feet 6 inches
New York, NY	5 feet 1 inch
Portland, ME	9 feet 11 inches

1. Make a graph to show the data in the table.

2. At which place is the difference between high and low tide the greatest? The least? _____

3. What is the range in these measurements? _____

4. Which two places have the closest differences? _____

5. Which place is the median among these measurements? _____

Mixed Strategy Review

6. Dov and his family arrive at the beach at 10:50 A.M. They leave to have lunch in a nearby restaurant at 1:10 P.M. and return to the beach at 2:00. Then they stay until 5:15. How long do they spend on the beach in all? _____

7. A day pass for the beach costs $5. If 65 people buy day passes, how much do they pay in all? _____

Spiral Review

Round to the nearest 10,000.

8. 4,600 _____   9. 5,213 _____   10. 136,000 _____

Name_____

 **6·9** **Multiply Using Mental Math**

Multiply. Use Mental Math.

1. $18 \times 41 =$ _____

2. $42 \times 29 =$ _____

3. $66 \times 28 =$ _____

4. $49 \times 12 =$ _____

5. $63 \times 21 =$ _____

6. $44 \times 37 =$ _____

7. $19 \times 50 =$ _____

8. $27 \times 26 =$ _____

9. $14 \times 48 =$ _____

10. $26 \times 32 =$ _____

11. $71 \times 17 =$ _____

12. $51 \times 28 =$ _____

13. $22 \times 20 =$ _____

14. $16 \times 55 =$ _____

15. $28 \times 28 =$ _____

Problem Solving

Use data from the table for problems 16–18.

16. If Boat A could travel at the same speed for 8 hours, how many miles could it travel?

_____

17. How much faster is Boat A than Boat G?

_____

18. What if Boat A and Boat G could travel at their top speeds for 8 hours, how much further would Boat A be than Boat G?

_____

Top Powerboat Speeds in Miles per Hour

Boat A	116
Boat B	115
Boat C	89
Boat D	80
Boat E	77
Boat F	73
Boat G	66

Spiral Review

19. 436
   $\times 94$

20. 7,998
   $+ 596$

21. 6,202
   $- 3,539$

22. 1,472
   $\times \ 69$

7-1 Division Patterns

Divide.

1. $4\overline{)80}$

2. $7\overline{)490}$

3. $5\overline{)\$350}$

4. $8\overline{)6,400}$

5. $9\overline{)\$7,200}$

6. $8\overline{)560}$

7. $210 \div 3 =$ _____

8. $3,600 \div 4 =$ _____

9. $\$500 \div 5 =$ _____

10. $24,000 \div 3 =$ _____

11. $42,000 \div 7 =$ _____

12. $54,000 \div 6 =$ _____

Problem Solving

13. Vijay finishes a math problem in 2 minutes. He has two hours to spend on math. How many math problems can he do? _____

14. Martha has a book to read for history class. She can read a page in 3 minutes. The book is 150 pages long. How long will it take her to finish it?

_____

Use data from the table for problems 15–16.

15. Janine likes hip-hop, but her brother likes jazz. If she buys two CDs of each type and takes the half-price offer, how much money does she need?

_____

16. Mei wants to buy 2 rock CDs while they are on sale. If she has four weeks to save up for her purchase, how much money does she have to save each week? _____

Big CD Sale

Buy one at sale price, get another for half price	
Hip-hop CDs	$10
Rock CDs	$8
Jazz CDs	$6

Spiral Review

17. $566 \times 20 =$ _____

18. $49 \times 40 =$ _____

19. $4 \times 60,000 =$ _____

20. $33 \times 58 =$ _____

Name_____

 **7·2** **Explore Division**

Divide.

1. 4)‾56‾   2. 7)‾98‾   3. 9)‾558‾   4. 8)‾265‾

5. 6)‾471‾   6. 3)‾448‾   7. 5)‾296‾   8. 3)‾551‾

9. 6)‾496‾   10. 4)‾252‾   11. 6)‾549‾   12. 9)‾842‾

13. 66 ÷ 5 = _____   14. 525 ÷ 4 = _____   15. 771 ÷ 9 = _____

16. 249 ÷ 6 = _____   17. 335 ÷ 7 = _____   18. 189 ÷ 3 = _____

19. 414 ÷ 9 = _____   20. 665 ÷ 7 = _____   21. 433 ÷ 6 = _____

Solve.

22. Josie solved a division problem. The divisor is 7. The quotient is 24. The remainder is 4. What is the dividend? _____

23. Josh collected cans for recycling. After he fills 6 bags that hold 26 cans each, he still has 4 cans left over. How many cans did Josh collect? _____

Spiral Review

Round to the nearest hundred.

24. 765 _____   25. 5,672 _____

26. 45,909 _____   27. 363,884 _____

Name_____

 **7·3** **Divide 3-Digit Numbers**

Divide. Check your answer.

1. 4)67   2. 6)98   3. 9)$558   4. 3)465

5. 6)774   6. 3)508   7. 2)$486   8. 3)202

9. 8)695   10. 5)352   11. 69 ÷ 7 = _____

12. 55 ÷ 3 = _____   13. 681 ÷ 7 = _____   14. $369 ÷ 9 = _____

15. 580 ÷ 7 = _____   16. 446 ÷ 9 = _____   17. 373 ÷ 8 = _____

18. 434 ÷ 8 = _____   19. 178 ÷ 6 = _____   20. 909 ÷ 6 = _____

Problem Solving

21. Mariko goes to Japanese school after regular school. Her class is making origami paper cranes. If the class makes 37 on Monday, 34 on Tuesday, and 35 on Wednesday, how many do they make in all on those three days?

_____

22. Mariko's class has made 555 paper cranes since they started. If there are 15 students in the class and each one made an equal number, how many paper cranes has each student made?

_____

23. If the class wants to make 1,000 paper cranes in all, about how many does each student still have to make?

_____

Spiral Review

24. 2,356
    + 344

25. 5,992
    + 4,339

26. 56,733
    + 41,376

© McGraw-Hill School Division

Name_____

7-4 Quotients with Zeros

Divide. Check your answer.

1. $5\overline{)57}$ 2. $8\overline{)89}$ 3. $5\overline{)545}$ 4. $3\overline{)315}$

5. $6\overline{)717}$ 6. $3\overline{)\$408}$ 7. $9\overline{)907}$ 8. $2\overline{)401}$

9. $6\overline{)819}$ 10. $4\overline{)425}$ 11. $663 \div 5 =$ _____

12. $405 \div 4$ _____   13. $971 \div 9 =$ _____   14. $\$240 \div 6 =$ _____

15. $\$735 \div 7 =$ _____   16. $\$189 \div 3 =$ _____   17. $526 \div 5 =$ _____

18. $830 \div 8 =$ _____   19. $405 \div 6 =$ _____   20. $709 \div 7 =$ _____

Find only those quotients that are greater than 600.

21. $4{,}664 \div 8 =$ _____   22. $4{,}664 \div 7 =$ _____   23. $5{,}199 \div 9 =$ _____

Problem Solving

24. Juan, Marvin, Kenji, and Abdullah together bowl 408. If each of them gets the same score, what does each one score?   _____

Use the data from the table for problems 25–26.

25. In the bowl-off, a team gets 2 stars for every point above 500 a game, and 3 points for every point above 550. How many stars do the Aliens get?

_____

26. If all 4 players on the Big Guys bowl the same score, what is each player's score?

_____

Team Bowling Scores

Aliens	565
Big Guys	492
Cool Captains	576
Diner Dudes	540

Spiral Review

27. $109 + 55 =$ _____   28. $507 - 249 =$ _____

29. $406 \times 22 =$ _____   30. $503 \times 19 =$ _____

Name_____

Interpret Remainders

Solve. Where appropriate, tell what inference you made and how you interpreted the remainder.

1. The Junior Service Club is planning a holiday party for the people at Shady Acres Nursing Home. They have a budget of $500. They put aside half for food and divide the rest equally among entertainment, decorations, and gifts. How much will they have for each?

   _____

   _____

2. The local Moose Club donates an extra $150, which the Juniors use for gifts. How much do they have for gifts now? How many $2 gifts can they buy?

   _____

3. The club decides to hire 4 acts: a singer, a dancer, a juggler, and a comedian. If all the performers are to get an equal share of the entertainment money, how much does the club pay each one?

   _____

Use data from the table for problems 4–5.

4. If the entertainment time is to be divided equally among the four acts, about how much time does each act get? _____

5. If it takes the same amount of time to get back to club HQ as it does to come from there, at what time should the members plan on getting back? _____

Shady Acres Party Schedule

10:00	Meet at club HQ
10:20	Arrive at Shady Acres to set up
12:00	Lunch
1:00	Entertainment
2:30	Gifts
2:45	Residents leave
3:00	Cleanup
4:00	Leave to return to club HQ

Spiral Review

6. $47 \times 33 =$ _____

7. $407 \times 33 =$ _____

8. $4,070 \times 33 =$ _____

9. $40,070 \times 33 =$ _____

Name_____

7·6 Estimate Quotients

Estimate. Choose compatible numbers.

1. 7)157　　　　2. 8)89　　　　3. 5)$508　　　　4. 3)615

5. 9)737　　　　6. 6)418　　　　7. 8)$506　　　　8. 4)211

9. 6)319　　　10. 9)425　　　11. $663 ÷ 9 = _____

12. 405 ÷ 7 _____　　　13. 471 ÷ 7 _____　　　14. $349 ÷ 6 _____

15. $635 ÷ 7 _____　　　16. $1,895 ÷ 3 _____　　　17. 526 ÷ 8 _____

18. 8,308 ÷ 9 _____　　　19. $4,085 ÷ 6 _____　　　20. 7,099 ÷ 8 _____

Problem Solving

21. Madison has $200 to buy 3 pairs of jeans and 3 tops. About how much does she have for each article of clothing?　　_____

22. Madison paid $32 each for 2 pairs of jeans, $23 for another pair of jeans, and $18 for each of 3 tops. She paid with 2 one-hundred-dollar bills. How much change does she get?　　_____

Spiral Review

23. 8,992 + 451 = _____　　　24. 8,992 − 451 = _____

25. 262 × 5 = _____　　　26. 262 ÷ 5 = _____

7·7 Divide 4-Digit Numbers

Divide. Check your answer.

1. $5\overline{)1,577}$ 2. $8\overline{)8,922}$ 3. $5\overline{)\$5,508}$ 4. $7\overline{)6,151}$

5. $9\overline{)7,637}$ 6. $6\overline{)5,418}$ 7. $4\overline{)\$5,066}$ 8. $2\overline{)2,911}$

9. $\$6,985 \div 5 =$ _____   10. $8,095 \div 8 =$ _____

11. $\$7,664 \div 9 =$ _____   12. $3,845 \div 3 =$ _____

13. $6,471 \div 7 =$ _____   14. $\$8,563 \div 4 =$ _____

15. $\$7,635 \div 7 =$ _____   16. $\$9,852 \div 3 =$ _____

17. $5,286 \div 8 =$ _____   18. $8,908 \div 2 =$ _____

19. Divide 9,002 by 7 _____   20. Divide $5,582 by 6 _____

Compare. Write < or >.

21. $4,664 \div 7$ _____ $4,166 \div 6$ 22. $8,525 \div 7$ _____ $9,064 \div 8$

23. $4,990 \div 5$ _____ $5,667 \div 6$

Problem Solving

24. A large box of wooden ice cream sticks contains 2,250 sticks evenly divided among 6 plastic bags. How many sticks are in each bag? _____

25. Nita makes 8 trays of sticks, using 216 sticks for each. How many sticks did she use? _____

Spiral Review

26. $689 - 343 =$ _____   27. $5,670 - 4,441 =$ _____

28. $54,990 - 14,056 =$ _____   29. $558,907 - 451,898 =$ _____

Name_____

7·8 Divide 5-Digit Numbers

Divide. Check your answer.

1. 7)34,176 **2.** 8)89,195 **3.** 4)$45,048 **4.** 3)11,615

5. 9)66,737 **6.** 6)15,418 **7.** 8)50,515 **8.** 6)21,719

9. 4)55,319 **10.** 9)14,025

11. 75,665 ÷ 9 = _____ **12.** 40,695 ÷ 6 = _____

13. 47,771 ÷ 3 = _____ **14.** 34,932 ÷ 7 = _____

15. 63,487 ÷ 8 = _____ **16.** 14,335 ÷ 4 = _____

17. 56,233 ÷ 6 = _____ **18.** 28,338 ÷ 9 = _____

19. 24,822 ÷ 2 = _____ **20.** 79,996 ÷ 5 = _____

Problem Solving

21. Attendance at the baseball stadium for the weekend is expected to be 99,400 for 4 games. If the same number of people attend each game, how many are there at one game? _____

22. Last year, attendance for the 4 games over the same weekend was 106,514. How many more people attended the 4 games than are expected this year? _____

Spiral Review

Find the range, median, and mode.

23. 17, 25, 19, 23, 27, 25, 33 _____

24. 450, 350, 375, 487, 337, 400, 450 _____

25. 36, 39, 39, 50, 24, 39, 41 _____

7·9 ▶ Find the Better Buy

Find each unit price. Find the better buy.

1. 4 ounces for $5.48
8 ounces for $9.92

_____

2. 3 hours for $12.00
5 hours for $20.00

_____

3. 2 pounds for $1.80
5 pounds for $4.70

_____

4. 3 gallons for $19.95
8 gallons for $40

_____

5. 3 yards for $159.90
9 yards for $530.10

_____

6. 5 cups for $9.80
8 cups for $13.76

_____

7. 4 quarts for $13.40
9 quarts for $33.93

_____

8. 2 pints for $5.70
5 pints for $15.00

_____

9. 5 inches for $15.50
9 inches for $26.01

_____

Problem Solving

Use data from the flyer for problems 10–12.

10. Bargain Barn is selling $\frac{1}{2}$-inch lace at $18 for 4 yards. Is it a better buy than the $\frac{1}{2}$-inch lace at Material Mart? _____

11. How much more does 4 yards of the most expensive trim cost than 4 yards of the least expensive trim?

_____

12. Silk trim costs $4.99 a yard. How much more does 5 yards of this cost than the same amount of $\frac{1}{2}$-inch lace? _____

Material Mart's Trim Sale!
Low Prices!

Material	Price per Yard
cotton piping	$1.19
$\frac{1}{4}$-inch lace	$3.49
$\frac{1}{2}$-inch lace	$4.49
satin	$2.29
velveteen	$2.99

Spiral Review

13. $5,000 \div 5 =$ _____

14. $72,000 \div 9 =$ _____

15. $320,000 \div 8 =$ _____

16. $1,500,000 \div 3 =$ _____

17. $4,900,000 \div 7 =$ _____

18. $350,000 \div 5 =$ _____

Name_____

Problem Solving: Strategy

Guess and Check

Use guess and check to solve.

1. Virgil's mother baked cookies for his scout meeting. There were 17 boys and 3 leaders there. She baked 120 cookies, and there were 23 left. If the leaders together ate 12 cookies, and each scout ate the same number of cookies, how many did each scout eat? _____

2. At the scouting awards night, 49 achievement arrows were given out. If Tian got 5 arrows and all the other scouts got at least 2, what is the largest number of scouts besides Tian who could have gotten more than 2? _____

3. Two of the 17 scouts are 8 years old. There are twice as many 10-year-old scouts as there are 9-year-old scouts. How many 9-year-old scouts are there? _____

Mixed Strategy Review

Use data from the table for problems 4–6.

Rosalind and her friends are playing a quiz game. Here are their scores:

4. Which player has the highest score? _____

5. Which players are tied for second-highest score?

_____

6. How many more points does the player in fourth place have than the player in last place?

_____

Angela	800
Annette	1,000
Darla	200
Karen	1,600
Rosalind	1,000

Spiral Review

Round to the nearest 1,000.

7. 802 _____

8. 46,337 _____

9. 355,900 _____

10. 4,550 _____

11. 99,958 _____

12. 37,615 _____

Name_____

 **7·11** **Explore Finding the Mean**

Find the mean.

1. 4, 7, 0, 3, 6 _____

2. 9, 12, 4, 16, 19 _____

3. 6, 10, 14 _____

4. 2, 7, 2, 3, 3, 8, 9, 6 _____

5. 3, 3, 9, 10, 5 _____

6. 4, 19, 3, 6, 8 _____

7. 4, 7, 8, 9 _____

8. 14, 19, 3, 8 _____

9. 7, 26, 1, 18 _____

10. 11, 13, 21 _____

11. 3, 3, 5, 6, 9, 4, 4, 6 _____

12. 6, 3, 8, 9, 2, 2, 2, 0 _____

13. 3, 1, 1, 1, 6, 9, 7 _____

14. 15, 3, 6, 7, 9 _____

15. 2, 2, 2, 2, 2, 3, 5, 6 _____

16. 5, 8, 7, 2, 3 _____

17. 2, 6, 6, 6, 8, 2 _____

18. 2, 9, 7, 8, 9, 1 _____

Solve.

19. Amy asked 6 friends how many phone calls they make each day. Here are their responses: 4, 4, 5, 2, 8, 7. What is the mean number of phone calls made by Amy's friends each day?_____

20. Eight students were asked how many states they have visited. Here are the answers they gave: 12, 17, 3, 6, 7, 6, 5, 8. What is the mean number of states visited by these students? _____

Spiral Review

Find the range, the median, and the mode.

21. 4, 4, 7, 9, 3 _____

22. 11, 10, 5, 3, 8, 8, 9 _____

23. 2, 7, 14, 14, 5 _____

24. 1, 5, 6, 9, 14, 15, 15 _____

Name_____

7·12 Find the Mean

Find the mean.

1. 15, 46, 9, 60, 35 _____

2. 17, 17, 17, 17, 17 _____

3. 566, 290, 330, 0, 14 _____

4. 718, 23, 98, 100, 56 _____

5. 48, 22, 35 _____

6. 51, 18, 13, 5, 2, 21, 19, 15 _____

7. 7, 24, 49, 9, 8, 62, 2 _____

8. 335, 42, 94 _____

9. 76, 110, 8, 9, 12, 13 _____

10. 101, 16, 16, 19 _____

11. 98, 6, 37, 212, 0, 475 _____

12. 13, 15, 14, 11, 5, 25, 22 _____

Problem Solving

Use data from the graph for problems 13–15.

13. What is the mean number of hours Nancy spent volunteering?

    _____

14. How many more hours would she have had to volunteer during the week for the mean to be 3?

    _____

15. What are the range, the median, and the mode for Nancy's schedule?

    _____

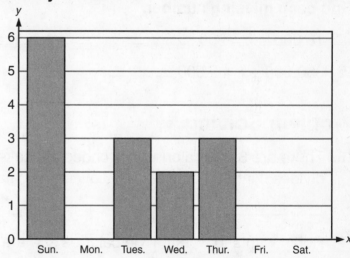

Nancy's Volunteer Hours at the Food Bank

Spiral Review

16. $422 \times 3 =$ _____

17. $1,015 - 47 =$ _____

18. $775 + 225 =$ _____

19. $12,345 \div 6 =$ _____

Name_____

8·1 Division Patterns

Write the number that makes each sentence true.

1. $18 \div 2 = a$ _____
2. $56 \div 7 = c$ _____
3. $15 \div e = 3$ _____

$180 \div 20 = a$ _____
$560 \div 70 = c$ _____
$f \div 50 = 3$ _____

$1,800 \div 20 = b$ _____
$5,600 \div 70 = d$ _____
$1,500 \div 50 = g$ _____

Divide. Use mental math.

4. $70\overline{)420}$
5. $80\overline{)720}$
6. $50\overline{)45,000}$
7. $30\overline{)1,200}$

8. $48,000 \div 60 =$ _____
9. $\$63,000 \div 70 =$ _____

10. $36,000 \div 60 =$ _____
11. $\$64,000 \div 80 =$ _____

12. $49,000 \div 70 =$ _____

Find each missing number.

13. $18,000 \div h = 900$ _____
14. $j \div 60 = 7$ _____

15. $360,000 \div k = 9,000$ _____

Problem Solving

16. There are 3,200 calories in a chocolate cake that Miranda baked. If she divides it into 8 equal pieces, how many calories does each piece have?

_____

17. Robin wants to work off the calories she ate in her piece of the chocolate cake. If she goes bicycling and works off 5 calories a minute, how long does she have to ride to work off all the calories?

_____

Spiral Review

18. $\$360 \times 5 =$ _____
19. $\$4,200 \times 5 =$ _____

20. $\$48,000 \times 5 =$ _____

Name_____

 **8·2** **Explore Dividing by 2-Digit Numbers**

Divide. You may use place-value models.

1. 24)96   **2.** 15)227   **3.** 16)84

4. 21)192   **5.** 11)223   **6.** 13)222

7. 19)312   **8.** 23)178   **9.** 14)208

10. 12)155

11. 88 ÷ 17 = _____   **12.** 133 ÷ 21 = _____

13. 154 ÷ 18 = _____   **14.** 176 ÷ 22 = _____

15. 259 ÷ 19 = _____   **16.** 230 ÷ 17 = _____

17. 278 ÷ 26 = _____   **18.** 244 ÷ 22 = _____

19. 119 ÷ 15 = _____   **20.** 166 ÷ 13 = _____

Solve.

21. The Martin family is moving to a new city, and Laurie is helping with the packing. There are 186 books on the bookshelf. Each box holds 24 books. How many boxes will Laurie fill? How many books will be left over?

_____

22. Taketo has set himself a goal of mastering 12 spelling words a day. The list from which he is learning contains 84 words. How many days will it take Taketo to learn all the words?

_____

Spiral Review

23. 6)14,336   **24.** 8)15,919

25. 16,700 ÷ 3 = _____   **26.** 20,551 ÷ 7 = _____

8·3 Divide 2-Digit Numbers by Multiples of 10

Divide.

1. 66 ÷ 30 = _____

2. 94 ÷ 10 = _____

3. 18 ÷ 10 = _____

4. 69 ÷ 20 = _____

5. 83 ÷ 50 = _____

6. 87 ÷ 40 = _____

7. 87 ÷ 20 = _____

8. 51 ÷ 30 = _____

9. 76 ÷ 20 = _____

10. 89 ÷ 30 = _____

11. 30)̄74

12. 50)̄71

13. 40)̄50

14. 30)̄60

15. 10)̄77

16. 20)̄93

Problem Solving

17. The Golden State Warriors basketball team play at the Oakland Arena, which seats 19,200 people. At one game, all but 1,244 seats were filled. How many people were there? _____

18. In the first 20 games of the season, a high school basketball player scores 99 points. About how many points does he average a game? _____

19. During a game, his team scores 59 points. About how many points does it score each quarter? _____

Spiral Review

20. 4)̄5,667

21. 6)̄45,972

22. 4)̄33,392

23. 2)̄51,666

24. 7)̄59,055

25. 8)̄32,144

Name _____

 **8·4** **Divide by 2-Digit Divisors**

Divide. Check your answer.

1. 23)90 **2.** 18)352 **3.** 27)656 **4.** 11)$89.87

5. $99.66 ÷ 33 = _____ **6.** 7,711 ÷ 24 = _____

7. 5,595 ÷ 49 = _____ **8.** $91.63 ÷ 17 = _____

9. 679 ÷ 44 = _____ **10.** $8.75 ÷ 35 = _____

Problem Solving

Use data from the scoreboard for problems 11–13.

After school, Mr. Saks, the gym teacher, took the entire 4th grade class to see a high school baseball game. The scoreboard shows how many runs each team made in each inning, plus the total runs, hits, and fielding errors. Since East High, the home team, was ahead in the 9th inning, it did not have to take its turn at bat.

11. In how many innings did West score more runs than East? The same number of runs? Fewer?

_____

_____

INNING											
1 2 3 4 5 6 7 8 9									**R**	**H**	**E**
West High											
1 0 4 3 0 0 0 2 0									10	14	5
East High											
0 0 2 2 6 1 0 0 X									11	16	2

12. What is the greatest number of runs by which one team outscored another in an inning?

_____

13. For each team, find the mean number of runs per inning, rounded to the nearest whole number.

_____

Spiral Review

14. 154 × 8 = _____ **15.** 254 × 8 = _____ **16.** 354 × 8 = _____

 **8·5 Estimate Quotients**

Estimate the quotient. Choose compatible numbers.

1. $47\overline{)189}$ **2.** $15\overline{)3,077}$ **3.** $41\overline{)2,920}$ **4.** $82\overline{)338}$ **5.** $63\overline{)1,471}$

_____   _____   _____   _____   _____

6. $789 \div 76$ _____   **7.** $623 \div 13$ _____

8. $1,023 \div 27$ _____   **9.** $2,223 \div 11$ _____

10. $4,905 \div 11$ _____   **11.** $4,232 \div 59$ _____

12. $1,167 \div 31$ _____   **13.** $2,290 \div 79$ _____

14. $6,789 \div 69$ _____

Problem Solving

15. Rahman has entered a walkathon for charity. The route is 26 miles long. If all the walkers finish, together they will walk 5,780 miles. About how many walkers are there?

_____

16. Omar practiced by walking 8 miles a day for 7 days. Rahman practiced by walking 10 miles on Monday, 14 on Tuesday, 8 on Wednesday, 11 on Thursday, and 14 on Saturday. Who walked more practice miles that week? How many more miles?

_____

Spiral Review

Estimate the product. Choose compatible numbers.

17. 29×87 _____   **18.** 46×23 _____

19. 77×34 _____   **20.** 39×61 _____

Name_____

8·6 Adjust the Quotient

Divide. Check your answer.

1. 22)196   2. 45)360   3. 36)200   4. 19)180   5. 34)279

6. 339 ÷ 41 = _____   7. 470 ÷ 67 = _____

8. 516 ÷ 82 = _____   9. 717 ÷ 98 = _____

10. $135 ÷ 27 = _____   11. $174 ÷ 29 = _____

12. 7,634 ÷ 53 = _____   13. $34.08 ÷ 24 = _____

Problem Solving

14. Every day after school Manuela goes to the gym to practice tumbling. She is supposed to do 275 forward and backward rolls a week. If she does 55 a day, will she do enough? Explain your answer. _____

Use data from the table for problems 15–16.

15. Lupe orders a sandwich, a side dish, and a soda that total the fewest calories. What does she order? How many calories do they have in all?

_____

_____

Calories in Fast Food	
Hamburger	409
Cheeseburger	525
Side of chili	315
Fish sandwich	389
Fries	276
Side salad	50
Regular soda	240
Diet soda	0

16. Max has to gain weight for wrestling. The coach has told him to eat about 1,000 calories for lunch. What combinations of sandwich, side dish, and soda when rounded to the nearest 100 come to 1,000?

_____

_____

Spiral Review

17. 45 ÷ 9 = _____   18. 565 ÷ 9 = _____   19. 5,733 ÷ 9 = _____

Name_____

Use an Overestimate or Underestimate

Solve. Tell why you used an overestimate or an underestimate.

1. An adult leader is needed for every 6 students on a camping trip. If there are 79 students, how many adults are needed?

   _____

2. On the first day of the trip, sandwiches are needed while the group is hiking to the campsite. Every student and leader is to get 2 sandwiches. Sandwiches come in cartons of 24. How many cartons should be ordered?

   _____

Use information from problems 1 and 2 and data from the table for problems 3–5.

Camping Supplies

Item	Amount Needed	Number Per Carton
Mess kits	94	12
Single-serving cereal	188	24
Loaves of bread	60	12

3. One night during the camping trip, the students are going to cook spaghetti. If a box serves 4, how many boxes have to be bought? Explain your answer. If boxes come 12 to a carton, how many cartons should be bought?

   _____

   _____

4. How many cartons of single-serving cereal have to be bought? _____

5. The leaders have set aside $500 for food. If they spend $16 on spaghetti, $60 for bread, and $47 for cereal, how much will they have left over for other things? _____

Spiral Review

Round.

6. 560 to the nearest 100 _____   7. 78 to the nearest 10 _____

Name_____

Problem Solving: Strategy

Choose a Strategy

Solve. Tell which strategy you used.

1. The track where Toshi runs is 100 yards around. If he runs around it 8 times without stopping, how many feet has he run? _____

2. The town is putting up new rows of seats beside the track. There will be 5 rows each 30 feet wide, and each seat will be 2 feet wide. How many seats will there be? _____

3. Towels for the locker room at the track come 12 to a package. Each folded towel is 4 inches high, and there are two equal stacks per package. How high is the package? _____

4. The announcement board at the track is 2 feet wide by 3 feet high. If it is divided into sections 6 inches wide by 6 inches high, how many sections can fit on the board? _____

5. In the summer Toshi runs three days a week. If summer vacation starts June 25 and ends September 3, how many days does he run during the summer? _____

6. If he begins his runs at 7:15 A.M. and ends at 8:00, then runs again from 7:15 to 8:00 P.M., how much does he run on a running day? _____

Mixed Strategy Review

7. A men's store restocking its shelves purchases 70 ties at $20 each, 2 boxes of bow ties at $240 a box, and 1 box of string ties at one-half the price of one box of bow ties. How much did the restocking cost? _____

8. Mario shoots baskets in his driveway every afternoon that the weather is nice. On Monday of one week he shot 50 baskets and made 15 of them; on Tuesday he shot 65 making 23; and on Thursday he made 13 baskets of 42. What was the average number of baskets he scored that week? _____

Spiral Review

9. $76 \div 30 =$ _____

10. $679 \div 40 =$ _____

11. $6,978 \div 20 =$ _____

12. $7,976 \div 20 =$ _____

8·9 Order of Operations

Write which operation should be done first.

1. $7 \times 8 \div 4$ _____

2. $9 + 4 \div 2$ _____

3. $(9 - 6) \times 5$ _____

4. $24 - 12 \div 4$ _____

5. $8 \times (3 + 5)$ _____

6. $8 \times 3 + 5$ _____

7. $16 - 5 + 7$ _____

8. $5 \times (14 \div 2)$ _____

Simplify. Use order of operations.

9. $6 + 5 - 2 =$ _____

10. $(14 - 5) \times 7 =$ _____

11. $(20 - 13) \times 9 =$ _____

12. $7 \times 9 - 25 =$ _____

13. $(8 \times 6) \div (2 \times 2) =$ _____

14. $45 \div (2 + 3) =$ _____

15. $(27 \div 3) \times (7 - 5) =$ _____

16. $(17 + 5) \div 2 =$ _____

Problem Solving

Use data from the table for problems 17–18.

17. Laszlo bought 4 pairs of sweat socks, 4 t-shirts, and 2 pairs of shorts. Write an expression to describe what he spent, then find the total.

_____

18. How much did Althea spend if she bought 3 of the most expensive item?

_____

Sale on Sports Clothing

T-shirts	$ 5
Shorts	$ 4
Tank tops	$ 3
Sweat socks	$ 2
Hooded sweatshirts	$10

Spiral Review

19. $454 \times 90 =$ _____

20. $708 \times 50 =$ _____

21. $880 \div 44 =$ _____

22. $960 \div 40 =$ _____

9·1 ▶ Explore Customary Length

Estimate and then measure. Tell what unit and tool you use.

1. The height of your desk _____

2. The width of your chair _____

3. The length of an automobile _____

4. The distance you can kick a soccer ball _____

5. How far you can stretch a rubber band _____

6. The height of a book _____

Choose the best estimate.

7. The distance around a suitcase

   A. 7 in.
   B. 7 ft
   C. 7 yd
   D. 7 mi _____

8. The width of a VCR

   A. 20 in.
   B. 20 ft
   C. 20 yd
   D. 20 mi _____

9. The distance from New York to Philadelphia

   A. 100 in.
   B. 100 ft
   C. 100 yd
   D. 100 mi _____

10. The width of a piano

    A. 5 in.
    B. 5 ft
    C. 5 yd
    D. 5 mi _____

Spiral Review

Tell how much time has elapsed.

11. 7:00 A.M. to 10:23 A.M. _____

12. 5:15 A.M. to 1: 20 P.M. _____

13. 9:55 A.M. to 3:00 A.M. _____

14. 6:07 P.M. to 4:14 A.M. _____

9-2 Customary Capacity and Weight

Which object holds more? Estimate the capacity for each.

1. A coffee cup or a wastepaper basket _____

2. A teapot or a soup bowl _____

Choose the best estimate.

3. A large pot

   A. 10 fl oz
   B. 10 c
   C. 10 qt
   D. 10 gal _____

4. A full tank of gasoline

   A 15 fl oz
   B 15 c
   C 15 qt
   D 15 gal _____

Problem Solving

5. Marta is having 6 friends over for lunch. She has a quart of milk in the house. Should she buy more? Explain your answer.

_____

_____

6. Marta's mother is buying food for Marta's lunch party. She figures on 1 oz of bologna and 1 oz of cheese on 2 slices of bread for each sandwich. She figures on 2 sandwiches a person. If she wants to buy the fewest number of separate packages and have the least left over, how many of what sizes should she buy?

At the Supermarket

Food	Package Sizes
American cheese	12 oz, 24 oz
Bologna	8 oz, 16 oz
Coleslaw	1 lb, 2 lb
Sliced bread	12, 18 slices

_____

Spiral Review

Which is the better buy?

7. 2 12-oz packages of cheese for $2.89 each or 1 24-oz package for $4.99? _____

_____

Name_____

 **9·3** **Convert Customary Units**

Write the number that makes each sentence true.

1. 16 qt = _____ gal
2. 36 in. = _____ ft
3. _____ c = 24 fl oz

4. 9 yd = _____ ft
5. 8 pt = _____ gal
6. _____ T = 10,000 lb

7. _____ lb = 96 oz
8. _____ ft = 96 in.
9. _____ gal = 96 c

10. 10 pt = _____ qt
11. 3 mi = _____ ft
12. 3 mi = _____ yd

13. _____ fl oz = 4 gal
14. 2 gal = _____ pt
15. 4 pt = _____ fl oz

16. 5 lb = _____ oz
17. 14 pt = _____ c
18. 112 oz = _____ lb

19. _____ fl oz = 3 pt
20. 48 fl oz = _____ c

Problem Solving

21. Lou Ann and her class visit a dairy farm. There are 25 cows, who together give an average of 17,500 lb of milk a month. How much does each cow give on average? _____

22. About how many tons of milk do the 25 cows give on average during a month? _____

23. About how many pounds of milk does a single cow give on average each day? Each year?

_____

24. The farmer is putting up a new fence around his pasture. On one side it measures 200 yd. How many ft is that? _____

Spiral Review

25. 56,334 + 2,789 = _____
26. 8,996 − 2,344 = _____

27. 665 × 23 = _____
28. 989 ÷ 54 = _____

9·4 **Problem Solving: Reading for Math**

Checking for Reasonableness

Solve. Explain your answer.

1. Carrie wants to put on a show in her family's empty barn. The barn is 30 yards long. She says that it is 60 feet long. Is her statement reasonable? Explain.

   _____

2. She says she can fit 10 rows of seats in the barn and still have plenty of room for a stage and a backstage area. If she allows 4 feet for each row, does her statement seem reasonable? Explain.

   _____

   _____

Use data from the list for problems 3–6.

3. Carrie says she can fit 20 seats in a row. Is her statement reasonable? Explain.

   _____

   _____

   _____

   _____

Carrie Measures the Barn

Width of barn	40 feet
Width of seat	24 inches
Width of center aisle	6 feet
Width of aisle between first row of seats and stage	6 feet
Stage from front to back	20 feet
Curtain height	7 feet

4. Carrie says that she can seat 200 people at a performance. Is her statement reasonable? Explain.

   _____

Spiral Review

5. If Carrie charges $5 a ticket and all the seats are sold, how much will she take in for a performance? _____

6. If a chair costs $2 to rent for one night, how much will Carrie have left if the show is sold out? _____

9·5 Explore Metric Length

Choose the best estimate.

1. The width of a four-lane street

   A. 10 cm
   B. 10 dm
   C. 10 m
   D. 100 m   _____

2. The distance from Chicago to Milwaukee

   A. 140 m
   B. 1,400 m
   C. 140 km
   D. 14,000 km   _____

3. The distance from your head to your toes

   A. 13 cm
   B. 130 cm
   C. 1,300 cm
   D. 13,000 cm   _____

4. The length of a straw

   A. 2 mm
   B. 2 cm
   C. 20 cm
   D. 200 cm   _____

Solve.

5. Matthew says he can throw a ball 1km. Is this reasonable? Why or why not?

   _____

6. Bari wants to frame a poster for her room. She says the distance around the poster is 5 cm. Is this reasonable? Why or why not?

   _____

Spiral Review

Tell how much time has elapsed.

11. 7:40 A.M. to 11:23 P.M. _____

12. 9:15 A.M. to 11:00 A.M. _____

13. 3:35 A.M. to 4:00 P.M. _____

14. 7:07 P.M. to 5:16 A.M. _____

Name_____

 **9-6** **Metric Capacity and Mass**

Estimate and then measure the capacity of each object. Order the objects from least to greatest capacity.

1. A juice box

2. A bathtub

3. A watering can

_____

Estimate and then measure the mass of each object. Order the objects from least to greatest mass.

4. A dictionary

5. A pencil

6. An orange

_____

Choose the best estimate.

7. A concert program
   10 g or 1 kg? _____

8. A trumpet
   10 g or 1 kg? _____

9. A paper cup
   100 mL or 1 L? _____

10. A box of peanuts
    6 g or 60 g? _____

Problem Solving

11. An usher at a concert carries a flashlight to help people find their seats. Does the flashlight have a mass of about 400 g or about 4 kg? _____

12. If the concert starts at 8:00, ends at 9:50, and there is a 20-minute intermission, how long is the actual concert? _____

Spiral Review

Find the product. Use mental math.

13. $4 \times 80 =$ _____

14. $40 \times 80 =$ _____

15. $40 \times 800 =$ _____

16. $400 \times 800 =$ _____

Name_____

9·7 ▷ Convert Metric Units

Write the number that makes each sentence true.

1. 10 L = _____ mL

2. 16 cm = _____ mm

3. _____ cm = 4 m

4. 12 m = _____ dm

5. 25,000 mL = _____ L

6. _____ g = 19 kg

7. _____ kg = 47,000 g

8. _____ m = 2,000 km

9. _____ L = 3,000 mL

10. 19 m = _____ mm

11. 45 m = _____ dm

12. 3 cm = _____ mm

13. _____ mL = 30 L

14. 2 km = _____ cm

15. 4 dm = _____ mm

16. 5 kg = _____ g

Compare. Write >, <, or =.

17. 40 cm _____ 40 dm

18. 750 L _____ 750 mL

19. 7,000 g _____ 7 kg

20. 300 dm _____ 30 m

Problem Solving

21. The recipe for the baker's largest cake calls for 2 kg of sugar. How many 400-gram packages should he use?   _____

22. Jib, one of the dancing poodles, has a mass of 6,000 grams. Rex, his brother, has a mass of 8 kg. Which dog weighs more? How much more?

_____

23. The dancing poodles' water bowl holds 1,400 mL of water. Is this more than, less than, or equal to 1 L?

_____

Spiral Review

24. 896 × 22 = _____

25. 7,549 × 34 = _____

26. 17,993 × 45 = _____

9·8 **Problem Solving: Strategy**

Logical Reasoning

Use logical reasoning to solve each problem.

1. Alice has only 3-, 4-, and 6-cup measures. At a party, she has to dish up 1-cup servings of chili to the children and 2-cup servings to the adults. How can she use the containers to measure 1 cup and 2 cups?

   _____

   _____

   _____

   _____

2. Alice has events to cater on four days next week. The Elks dinner is 2 days after the Rotary meeting. The Morgan party is 6 days before Angela's sweet sixteen party, which falls on Saturday two days after the Elks dinner. On which days are the other events?

   _____

   _____

3. At the Elks dinner, there is a choice of 4 main courses: eggplant, fish, chicken, and beef. Four times more people order beef than chicken. Twice as many people order chicken as fish. Six times more people order fish than eggplant, which has just 3 orders. How many people order the other choices? How many people are at the dinner?

   _____

Mixed Strategy Review

4. Alice's largest pot holds 12 quarts. The Elks have one that holds 6 gallons. Is this 2, 3, or 4 times larger than Alice's pot? _____

5. The Elks have 146 cups, but they need 201. If they buy 8 cups a week, how many weeks will it take them to get the cups they need? _____

Spiral Review

6. 10 kg = _____ g   7. 90 L = _____ mL   8. 40 km = _____ m

Name_____

 **9·9** ▶ **Temperature**

Give a reasonable temperature for each.

1. Cold lemonade _____

2. A warm bath _____

Choose the best estimate.

3. A nice day for a baseball game

   22° F or 22° C _____

4. A good day for ice skating

   28° F or 28° C _____

Problem Solving

5. Suppose it is 66° F during the day and it drops 19° during the night. What temperature is it during the night?

_____

6. Suppose it is 12° C during the day and it drops 14° during the night. What temperature is it during the night?

_____

Use data from the bar graph for problems 7–8.

7. How many degrees difference is there between the highest high temperature and the lowest high temperature?

_____

8. Akira wants to attend a baseball game on Wednesday. Should he wear a jacket? Why?

_____

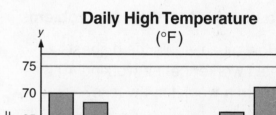

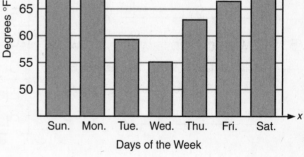

Daily High Temperature
(°F)

Spiral Review

9. 8 ft = _____ in.

10. 2 mi = _____ yd

11. 8 lb = _____ oz

 **10·1** # 3-Dimensional Figures

Identify the 3-dimensional figure the object looks like. Tell how many faces or bases, edges, and vertices it has.

1.

_____

Identify the figures used to construct this building.

2.

_____

Copy and fold. Identify the 3-dimensional shape.

3.

_____

Use data from the chart for problems 4–5.

4. How much would the highest-paid worker get for working 40 hours a week for 4 weeks?

_____

5. How much more was a drywall installer paid than a plasterer for working 40 hours a week?

_____

Selected Construction Workers' Salaries, 1998

Occupation	Mean Hourly Salary
Brick mason	$17.81
Carpenter	$15.20
Drywall installer	$15.50
Electrician	$18.05
Plasterer	$15.34

Spiral Review

6. $389 \times 59 =$ _____

7. $389 \div 59 =$ _____

8. $389 + 59 =$ _____

Name_____

Tell whether each figure is open or closed. Is it a polygon? If so, classify the figure.

1. _____

_____

2. U _____

_____

Draw the figure and identify it.

3. a 6-sided polygon _____

4. a closed figure that has one straight line _____

5. a 4-sided figure _____

Problem Solving

6. A floor has a design of triangles. There are 40 rows of triangles, with 50 triangles in each row. How many triangles does the floor design have?

_____

Use data from the picture for problem 7.

7. How many rectangles are in the drawing? Squares? Triangles?

_____

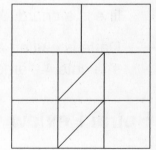

8. Name a polygon you could make by putting two squares together. Draw and label the figure.

_____

Spiral Review

9. 600 cm = _____ m   10. 12,000 lb = _____ T   11. 12 pt = _____ qt

Name_____

Identify each figure.

1.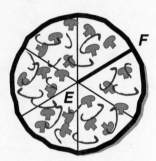

_____

_____

_____

2.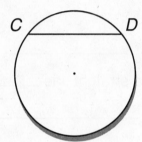

_____

_____

_____

Identify the parts of a circle.

3.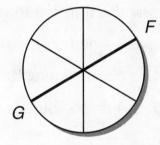

_____

4.

_____

5.

_____

Problem Solving

6. Tim and Greg are talking about today's lesson. They cannot agree about the longest straight line in a circle. Tim says that the diameter is longest. Greg thinks that the longest straight line is a chord. Which boy is right? _____

7. Darla is cutting a large pizza pie to share with her family. If she cuts 4 diameters, how many pieces will she get? _____

Spiral Review

9. $5,909 + 3,454 + 636 =$ _____

10. $5 \times 8 \times 9 =$ _____

11. $7,767 - 5,789 =$ _____

 **10·4** **Angles**

Write *acute, obtuse, or right* for each angle. Where there is more than one angle in a figure, write the angle for each letter.

1.

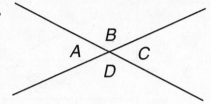

_____

2.

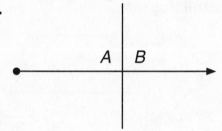

_____

Write the degree measure and fraction of a turn that each time suggests.

3.

4.

5.

_____   _____   _____

Problem Solving

6. Draw a polygon with twice as many sides as a

triangle. Label it._____

7. Draw a polygon with twice as many sides as a

square. Label it. _____

Spiral Review

8. 585 ÷ 24 = _____   **9.** 675 ÷ 22 = _____   **10.** 8,866 − 5992 = _____

Name_____

 **10·5** **Triangles and Quadrilaterals**

Classify each triangle as *equilateral, isosceles,* or *scalene.* Then classify each triangle as *right, acute,* or *obtuse.*

1.

_____

2.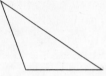

_____

3.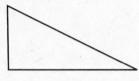

_____

Identify each quadrilateral. Then say if it is a parallelogram.

4.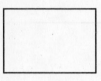

_____

5.

_____

6.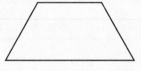

_____

Tell if the statement is true or false. Explain why.

7. All scalene triangles are obtuse.

_____

Problem Solving

8. The block that Ramona and Tricia live on is 400 feet long and 400 feet wide. Ramona says the block is a rhombus, and Tricia says that it is a square. With the information given, who do we know is right? What more information would be helpful? _____

_____

9. Warren says that squares and equilateral triangles are very much alike. Why would he say that? _____

Spiral Review

Write >, <, or =.

10. 300 cm _____ 3 m   **11.** 16 pt _____ 4 qt   **12.** 1,650 yd _____ 1 mi

10·6 **Problem Solving: Reading for Math**

Use a Diagram

Use the diagrams to solve each problem.

1. Della drew this polygon. Describe it in as many ways as you can.

_____

_____

2. Della wants to attach a triangle to the side of this figure, as shown. If the two sides of her triangle are 4 and 6 cm, is it possible that the third side will be 12 cm? Explain.

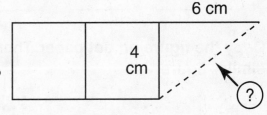

_____

3. Describe all the polygons in this figure that Roger drew.

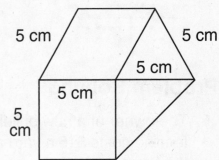

_____

_____

_____

4. What side length do all these polygons have in common? _____

5. Which of these is a reasonable estimate for the third side of the isosceles triangle: 4 cm, 7 cm, or 11 cm? _____

6. If Roger wants to add another equilateral triangle next to the existing one, what figure will the two of them make together? _____

Spiral Review

7. $9 \times 4 \times 3 =$ _____

8. $10 \times 4 \times 8 =$ _____

9. $8 \times 8 \times 7 =$ _____

Name_____

 **10·7** **Congruent and Similar**

Write whether the figures are similar. Then write whether the figures are congruent.

1. 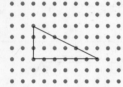 _____

2. 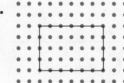 _____

Copy the figure on dot paper. Then draw one congruent figure and one similar figure.

3. **4.** **5.**

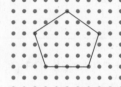

Problem Solving

6. The owner of a new building decides to put up a fence around the property. If fencing costs $15 a yard and she needs 389 feet of fencing, how much will the fence cost? If she has budgeted $2,000 for the fence, will she have enough? How much will she have left over for a gate?

_____

Use data from the chart for the problem.

7. Which row shows figures that are neither congruent nor similar to the first figure in the row?

_____

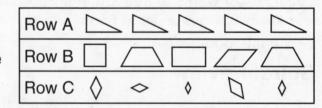

Spiral Review

8. $80.52 × 12 = _____   **9.** $966.24 × 30 = _____

© McGraw-Hill School Division

Name_____

 **10·8** **Explore Translations, Reflections, and Rotations**

Write *translation, reflection,* or *rotation* to describe how the figure was moved.

1.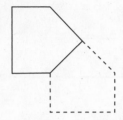

2.

3.

_____   _____   _____

Solve.

4. The Martinez family has moved to a new apartment and has ordered wall-to-wall carpeting for the bedrooms. The master bedroom is 14ft × 17ft. Rosa's room is 9ft × 12ft. Carpeting costs $5.80 per ft². How much more expensive is it to carpet the master bedroom?

_____

5. The blocks that make up a sidewalk are 6ft × 12ft. The blocks for a driveway are 6ft × 6ft. What is the difference between the area of a sidewalk block and a driveway block? The difference between their perimeters?

_____

Spiral Review

Write *true* or *false* for each of the statements in problems 6–9.

6. An equilateral triangle may have a right angle. _____

7. An isosceles triangle may have a right angle. _____

8. A scalene triangle may have a right angle. _____

9. A rectangle may be a rhombus. _____

Name_____

 **10·9** **Symmetry**

Is the dotted line a line of symmetry?

1.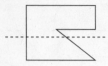

_____

2.

_____

3.

_____

Is the figure symmetrical? If yes, draw its line of symmetry.

4.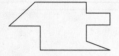

_____

5.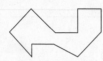

_____

6.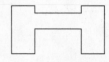

_____

7. Draw a figure with rotational symmetry.

Problem Solving

8. A walnut bookcase costs $860.59 at Quality Imports. A similar bookcase costs $795.89 at Main Street Cabinets. Which is more expensive? What is the difference in price?

_____

9. Janie and her friends like to play hopscotch at recess. Is the hopscotch board symetrical? If so, what kind of symmetry does it represent?

_____

Spiral Review

Order from greatest to least.

10. 10 m, 5 dm, 100 cm _____

11. 40 pt, 4 gal, 24 qt _____

12. 1,500 yd, 1 mi, 4,000 ft _____

13. 5,000 oz, 4 lb, 1 T _____

© McGraw-Hill School Division

Name_____

10·10 **Problem Solving: Strategy**

Find a Pattern

Use data from this tessellation for problems 1–2.

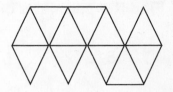

1. What shapes do you see in a repeated pattern? How are the figures moved?

   _____

2. Copy the pattern and add the missing pieces.

Use data from this tessellation for problems 3–4.

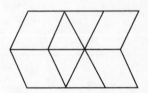

3. Describe the pattern using shapes, and tell how the figures are moved.

   _____

   _____

4. Copy the pattern and add the next 6 shapes.

Mixed Strategy Review

5. Barry's father is going to repair the driveway. He has $100 to spend. He buys a tub of driveway patch for $15.95, a pair of driveway asphalt for $44.95, 2 pairs of work gloves for $1.98 each, a putty knife and a trowel for $3.98 each, and a broom for $13. Does he go over his budget? If not, how much is left over?

   _____

6. The inside of Marla's house measures 21 feet by 54 feet. What are the dimensions in yards?

   _____

Spiral Review

7. 6ft =_____yd

8. 4 gal =_____pt

9. 68oz = ____gal____oz

10. 17yd =_____in.

11. 516 = _____oz

12. 18in. = _____ft

Name_____

 **Perimeter**

Find the perimeter of each figure.

1.
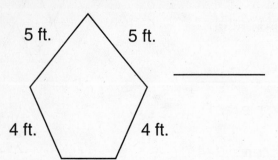
5 ft.   5 ft.

4 ft.   4 ft.

3 ft.

_____

2.   5 ft.

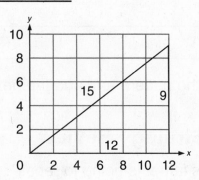

_____

3. _____

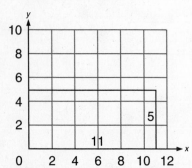

4. _____

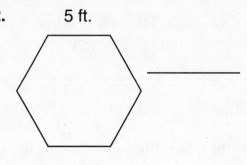

Use data from the plans for problems 5–7.

5. What is the perimeter of bedroom 1?

_____

6. Convert your answer in problem 5 to yards.

_____

7. Convert your answer in problem 6 to inches.

_____

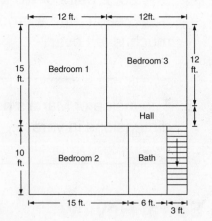

Spiral Review

Write the following spelled-out numbers in numerals.

8. four million, six hundred thousand, four _____

9. eleven million, four hundred sixty-six thousand _____

Name_____

 **Area**

Find the area of each figure.

1. 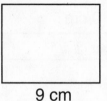 7 cm 9 cm _____

2.  8 cm _____

Use the grid to draw the figure.
Tell what the figure is and find the area.

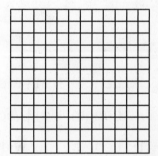

3. Length: 9 cm; width: 6 cm

_____

Find the area and perimeter of each figure.

4. 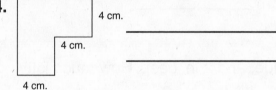 4 cm. 4 cm. 4 cm.

_____

_____

5. 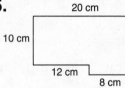 20 cm 10 cm 12 cm 8 cm

_____

_____

Problem Solving

6. A farmer wants to put a fence around a square field. The field is 300 ft on a side. How many yards of fencing does he need?

_____

7. The farmer must fertilize the square field. Fertilizer is sold in bags that cover 10,000sq ft. How many bags of fertilizer does he need?

_____

Spiral Review

8. $4 + 3 \times 20 =$ _____

9. $(4 + 3) \times 20 =$ _____

10. $4 \times 3 + 20 =$ _____

Name_____

Find the volume of each rectangular prism.

1.
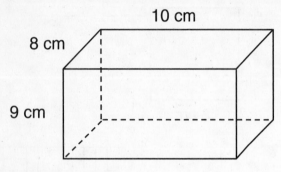
10 cm
8 cm
9 cm

_____

2.
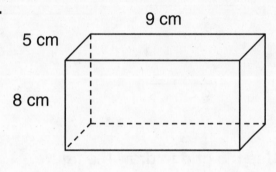
9 cm
5 cm
8 cm

_____

3. Length: 5 in.; width: 2 in.; height: 6 in. _____

4. Length: 8 in.; width: 2 in.; height: 4 in. _____

5. Length: 7 ft.; width: 5 ft.; height: 8 ft _____

6. Length: 17 cm.; width: 4 cm; height: 6 cm _____

Solve.

7. The kitty litter box is 17in. long, 9 in. wide, and 2 in. deep. How much kitty litter does it hold?

_____

8. Annette is pouring sugar from a 5 lb bag into a cannister 6in. long, 4in. wide, and 10in. deep. How much sugar can she pour into a cannister?

_____

Spiral Review

Find the area.

9. A rectangle 8 cm × 5 cm _____

10. A square that has a perimeter of 16 cm _____

11. A rectangle 100 yd × 50 yd _____

12. A square with a side of 18 ft _____

Name_____

 **11·1** ▶ **Parts of a Whole**

Write a fraction for the part that is shaded.

1.  ___

2. 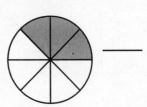 ___

3. 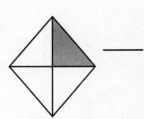 ___

Draw a rectangle with the fraction shaded.

4. $\dfrac{5}{6}$

5. $\dfrac{2}{5}$

6. $\dfrac{3}{5}$

7. $\dfrac{4}{9}$

8. $\dfrac{3}{7}$

9. $\dfrac{3}{8}$

Problem Solving

10. The roller coaster in Space Land is 184 feet high. The one in Monster Land is 235 feet high. How much higher is the one in Monster Land? _____

11. What fraction of the height of the Monster Land roller coaster is the Space Land roller coaster? _____

12. The Space Land roller coaster has 100 cars. Of these, 23 cars are red. What fraction of the cars are not red? _____

Spiral Review

13. $780 + 155 =$ _____

14. $780 - 135 =$ _____

15. $4,000 \times 20 =$ _____

16. $4,000 \div 20 =$ _____

 **II-2** **Parts of a Group**

Write a fraction that names what part is squares.

1. 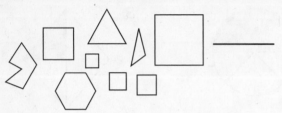 _____

2.  _____

3.  _____

4. 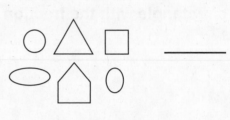 _____

Draw a picture, then write a fraction.

5. Three of seven pails have handles. _____

6. Two of five hot dogs are in buns _____

7. Seven of eight balloons have strings. _____

Problem Solving

8. Waiting time for the Space Land roller coaster is 20 minutes. What fraction of an hour is that? _____

9. At Space Land, Beth buys and uses 11 ride tickets. If there are 37 rides in Space Land, what fraction of them does Beth miss? _____

Spiral Review

10. $75\overline{)571}$ 11. $75\overline{)5,719}$ 12. $75\overline{)57,196}$

13. $456 \times 23 = $ _____   14. $4,566 \times 23 = $ _____

Name_____

 **11·3** **Find Equivalent Fractions and Fractions in Simplest Form**

Draw an equivalent fraction for each.

1.

2.

Complete to find equivalent fractions.

3. $\dfrac{3 \times 2}{6 \times \square} = \dfrac{6}{\square}$

4. $\dfrac{9 \div 3}{12 \div \square} = \dfrac{3}{\square}$

5. $\dfrac{7}{8} = \dfrac{14}{\square}$

Name an equivalent fraction for each.

6. $\dfrac{22}{24}$ _____

7. $\dfrac{7}{9}$ _____

8. $\dfrac{16}{64}$ _____

9. $\dfrac{35}{42}$ _____

Write each fraction in simplest form.

10. $\dfrac{50}{75}$ _____

11. $\dfrac{55}{60}$ _____

12. $\dfrac{27}{36}$ _____

13. $\dfrac{12}{72}$ _____

Complete the pattern of equivalent fractions.

14. $\dfrac{1}{3} = \dfrac{\square}{6} = \dfrac{\square}{9} = \dfrac{6}{\square} = \dfrac{8}{\square}$

15. $\dfrac{2}{5} = \dfrac{\square}{10} = \dfrac{\square}{15} = \dfrac{8}{\square} = \dfrac{10}{\square}$

Problem Solving

16. Mikhail's class voted for their favorite ride at Space Land. Of 24 students, 18 voted for the roller coaster. What fraction shows how many voted for the roller coaster? Write your answer in simplest form. _____

17. Of 27 students in Ms. Morey's class, 15 voted for the ferris wheel as their favorite ride. What fraction shows how many voted for the ferris wheel? Write your answer in simplest form. _____

Spiral Review

18. A rectangle is 12 cm long and 4 cm wide. What is its area? _____

19. What is the perimeter of the rectangle in problem 18? _____

Name_____

Compare. Write >, <, or =.

1. $\dfrac{3}{4}$ _____ $\dfrac{3}{5}$

2. $\dfrac{5}{6}$ _____ $\dfrac{10}{12}$

3. $\dfrac{5}{7}$ _____ $\dfrac{7}{14}$

Order from least to greatest.

4. $\dfrac{3}{4}, \dfrac{1}{2}, \dfrac{1}{4}$ _____

5. $\dfrac{5}{9}, \dfrac{2}{9}, \dfrac{1}{3}$ _____

6. $\dfrac{7}{10}, \dfrac{3}{10}, \dfrac{2}{5}$ _____

Order from greatest to least.

7. $\dfrac{5}{6}, \dfrac{1}{6}, \dfrac{1}{3}$ _____

8. $\dfrac{1}{8}, \dfrac{5}{8}, \dfrac{3}{4}$ _____

9. $\dfrac{2}{5}, \dfrac{3}{10}, \dfrac{3}{5}$ _____

Problem Solving

10. At Space Land, Jenna has $50 to spend on things other than admission for herself and her two cousins. If she spends $\dfrac{2}{5}$ of the money on Thomas and $\dfrac{3}{10}$ of the money on Iris, what fraction of the money is left for her? _____

Use data from the graph for problems 11–12.

11. Which food do more people buy than any other?

_____

12. If $\dfrac{2}{5}$ of the foods sold are hamburgers, what fraction are not hamburgers? _____

**Space Land
Foods People Buy**

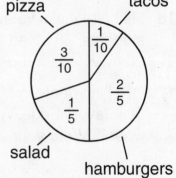

Spiral Review

Compare. Write >, <, or =.

13. 50,898 _____ 50,989

14. 508,889 _____ 508,889

15. 509,998 _____ 509,099

16. 510,898 _____ 501,898

 **II·5** **Problem Solving: Reading for Math**

Check for Reasonableness

Answer each question. Check for reasonableness.

1. At Space Land, a group of 12 students rode the Comet Trail ride. Of the group, $\frac{1}{3}$ wanted to take the ride again. How many students wanted to take the ride again?   _____

2. A group of 10 students went to the Mars! Mars! Mars! ride. There was room for only 7. What fraction of the group had to wait for the next trip?   _____

Use data from the table for problems 3-5.

The table shows the number of tickets each student used for rides during a day at Space Land.

3. Of the 16 tickets he used, Paul used 4 for the Star Flight ride. What fraction did he use for Star Flight? _____

4. Of the 10 tickets she used, Olivia used $\frac{2}{5}$ for the Galaxy Journey ride. How many tickets did she use for Galaxy Journey?

   _____

5. Of all the tickets used by the five students, what fraction did Felice use?

   _____

Space Land Tickets Used

Student	Tickets
Kareem	14
Felice	12
Olivia	10
Paul	16
Leanne	8

Spiral Review

Compare. Write >, <, or =.

6. $\frac{6}{12}$ ____ $\frac{5}{10}$ 7. $\frac{9}{11}$ ____ $\frac{17}{22}$ 8. $\frac{6}{7}$ ____ $\frac{11}{14}$ 9. $\frac{3}{4}$ ____ $\frac{7}{8}$ 10. $\frac{12}{13}$ ____ $\frac{24}{26}$

Name_____

 **11·6 Explore Finding Parts of a Group**

Find the fraction of each number.

1. $\frac{1}{7}$ of 49 _____   2. $\frac{3}{5}$ of 20 _____   3. $\frac{1}{6}$ of 36 _____   4. $\frac{1}{4}$ of 36 _____

5. $\frac{1}{9}$ of 18 _____   6. $\frac{3}{7}$ of 21 _____   7. $\frac{2}{3}$ of 33 _____   8. $\frac{3}{10}$ of 40 _____

9. $\frac{3}{8}$ of 32 _____   10. $\frac{2}{7}$ of 14 _____   11. $\frac{1}{5}$ of 30 _____   12. $\frac{1}{6}$ of 30 _____

13. $\frac{1}{8}$ of 8 _____   14. $\frac{1}{4}$ of 48 _____   15. $\frac{1}{3}$ of 39 _____   16. $\frac{1}{7}$ of 35 _____

Solve.

17. Of the 30 rides at Space Land, $\frac{3}{5}$ cost more than 1 ticket. How many rides cost more than 1 ticket?   _____

18. Of the 24 food stands at Space Land, 16 sell hot dogs. What fraction sell hot dogs?   _____

Spiral Review

Name the figure.

19.
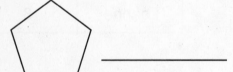   _____

20.
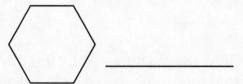   _____

21.
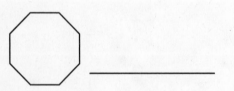   _____

22.
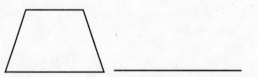   _____

Name_____

Rename each as a mixed number or fraction in simplest form.

1. $\frac{14}{5}$ _____

2. $\frac{27}{8}$ _____

3. $\frac{49}{12}$ _____

4. $\frac{67}{10}$ _____

5. $\frac{56}{10}$ _____

6. $\frac{11}{2}$ _____

Use the number line to compare. Write >, <, or =.

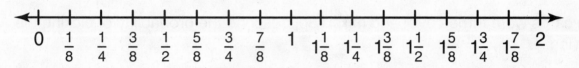

7. $1\frac{1}{2}$ _____ $1\frac{5}{8}$

8. $1\frac{1}{4}$ _____ $\frac{7}{8}$

9. $1\frac{3}{8}$ _____ $1\frac{3}{4}$

10. 1 _____ $\frac{8}{8}$

11. $1\frac{1}{2}$ _____ $\frac{1}{2}$

12. $1\frac{1}{4}$ _____ $1\frac{1}{8}$

Problem Solving

13. Colette had to wait $1\frac{2}{3}$ hours to get on the Galaxy Journey

    ride. Santosh had to wait $1\frac{5}{6}$ hours. Who waited longer? _____

14. At the end of the day, Sergiou figured he had waited a total

    of $2\frac{3}{8}$ hours for rides. Manny figured he had waited $2\frac{1}{2}$ hours.

    Who had waited longer? _____

Spiral Review

Order from least to greatest.

15. $\frac{1}{3}, \frac{5}{6}, \frac{2}{3}$ _____

16. $\frac{4}{5}, \frac{1}{5}, \frac{6}{10}$ _____

17. $\frac{5}{7}, \frac{6}{7}, \frac{9}{21}$ _____

Name_____

 **11·8 Likely and Unlikely**

Use the words *likely, equally likely, most likely, certain, unlikely,* or *impossible* to describe the probability of spinning the following:

1. 1, 2, or 3 _____

2. 1 _____

3. 3 _____

If the cards were turned over and mixed up, describe the probability of picking the following:

4. A _____   **5.** D _____   **6.** E _____

Describe the probability.

7. The roller coaster at Space Land will pick up speed when it goes downhill.   _____

8. The Star Flight ride will actually take off into space.   _____

Problem Solving

9. An acre takes up 43,560 square feet. If Space Land is 52 acres, how many square feet is that?   _____

10. If $\frac{1}{4}$ of Space Land is taken up by the Red section, how many acres is that?   _____

Spiral Review

Write each fraction in its simplest form.

11. $\frac{4}{32}$ ___   **12.** $\frac{5}{35}$ ___   **13.** $\frac{6}{36}$ ___   **14.** $\frac{7}{35}$ ___

15. $\frac{12}{16}$ ___   **16.** $\frac{6}{24}$ ___   **17.** $\frac{9}{27}$ ___   **18.** $\frac{12}{20}$ ___

Name_____

 **11·9** **Explore Probability**

If the cards shown above are mixed up and placed into a bag, find the probability of picking each shape.

1. Circle _____

2. Rectangle or star _____

3. Trapezoid _____

4. Trapezoid or star _____

5. Rhombus _____

6. Square _____

There are 5 red disks, 4 blue disks, 3 yellow disks, 2 green disks, and 1 purple disk in a bag. Find the probability of picking the color from the bag.

7. Red _____

8. Purple _____

9. Yellow or green _____

10. Yellow or purple _____

Solve.

11. There are 2 yellow marbles, 3 black marbles, 6 green marbles, and 1 white marble in a bag. If you picked one without looking, which color are you most likely to pick? Explain.

_____

12. Which color are you least likely to pick? Explain.

_____

Spiral Review

Rename each as a mixed number or fraction in simplest form.

13. $\frac{22}{7}$ _____

14. $\frac{24}{9}$ _____

15. $\frac{14}{4}$ _____

Name_____

Draw a Tree Diagram

Make a tree diagram to solve.

1. At Space Land, Jomo has to decide the order in which he
   goes on the following rides: Asteroid Attack (A), Comet Trail (C),
   Joltin' Jetstream (J), and Mars! Mars! Mars! (M). How many
   different orders are there?   _____

Tree diagram:

Mixed Strategy Review

2. If you toss two 1–to–6 number cubes, how many possible combinations could
   they make that contain a 1? Write the combinations. (Count the same
   combination from the two different cubes as two different combinations, for
   example: 2 + 4 and 4 + 2 count as 2 combinations.)

   _____

Use data from the table for problem 3.

3. Kees has to stack supplies for the food stand.
   He cannot place more than 100 pounds on each
   of 3 shelves. What items should he put where?

Buns and rolls	58 lb
Canned goods	96 lb
Cartons and trays	51 lb
Crunchy snacks	43 lb
Desserts	40 lb

   _____

   _____

   _____

Spiral Review

58, 96, 51, 62, 43, 23, 40, 15

4. Find the range. _____   5. Find the mean. _____

Name_____

 **11·11** ▶ **Explore Making Predictions**

There is a spinner with 5 equal sections: red, blue, yellow, green, purple.

1. If you spin the spinner 100 times, what is the probability that you will land on yellow?

   _____

2. If you spin the spinner 10 times, what is the probability that you will land on red or blue?

   _____

3. If you spin the spinner 50 times, what is the probability that you will land on green or purple?

   _____

There is a bag with the following cubes: red, 24; blue, 18; yellow, 3; green, 10; purple, 5.

4. If you pick a cube out of the bag 60 times, what is the probability that you will pick a blue cube?

   _____

5. If you pick a cube out of the bag 30 times, what is the probability that you will pick a red cube?

   _____

6. If you pick a cube out of the bag 60 times, what is the probability it will be green or purple?

   _____

Spiral Review

Draw:

7. an equilateral triangle

8. a parallelogram

9. a pentagon

10. a circle with a chord

Name_____

 **12-1** ## Add Fractions with Like Denominators

Add. Write each sum in simplest form.

1. $\dfrac{4}{7}$
    $+\dfrac{5}{7}$

2. $\dfrac{3}{4}$
    $+\dfrac{3}{4}$

3. $\dfrac{2}{5}$
    $+\dfrac{4}{5}$

4. $\dfrac{1}{8}$
    $+\dfrac{7}{8}$

5. $\dfrac{15}{16}$
    $+\dfrac{7}{16}$

6. $\dfrac{3}{10}$
    $+\dfrac{7}{10}$

7. $\dfrac{4}{7}$
    $+\dfrac{1}{7}$

8. $\dfrac{5}{12}$
    $+\dfrac{11}{12}$

9. $\dfrac{13}{18}$
    $+\dfrac{5}{18}$

10. $\dfrac{9}{10}$
    $+\dfrac{7}{10}$

11. $\dfrac{1}{9} + \dfrac{1}{9} =$ _____

12. $\dfrac{3}{17} + \dfrac{7}{17} =$ _____

13. $\dfrac{5}{6} + \dfrac{5}{6} =$ _____

14. $\dfrac{3}{11} + \dfrac{5}{11} =$ _____

15. $\dfrac{7}{22} + \dfrac{9}{22} =$ _____

16. $\dfrac{5}{16} + \dfrac{7}{16} =$ _____

Problem Solving

17. Carol's family uses $\dfrac{5}{8}$ of a quart of milk on Monday and another $\dfrac{5}{8}$ of a quart on Tuesday. How much milk do they use in all? _____

18. At the florist's shop, $\dfrac{7}{10}$ of the roses are red and $\dfrac{1}{10}$ of the roses are yellow. What fraction of the roses are either red or yellow? _____

Spiral Review

Order from the least to the greatest.

29, $\dfrac{7}{12}, \dfrac{3}{12}, \dfrac{5}{12}$ _____

21. $\dfrac{1}{10}, \dfrac{3}{5}, \dfrac{7}{10}$ _____

22. $\dfrac{7}{8}, \dfrac{5}{8}, \dfrac{3}{4}$ _____

Name_____

 **12·2 Subtract Fractions with Like Denominators**

Subtract. Write each difference in simplest form.

1. $\dfrac{7}{9}$
    $-\dfrac{4}{9}$

2. $\dfrac{11}{12}$
    $-\dfrac{7}{12}$

3. $\dfrac{5}{8}$
    $-\dfrac{1}{8}$

4. $\dfrac{9}{10}$
    $-\dfrac{1}{10}$

5. $\dfrac{17}{19}$
    $-\dfrac{14}{19}$

6. $\dfrac{4}{15}$
    $-\dfrac{2}{15}$

7. $\dfrac{13}{18}$
    $-\dfrac{5}{18}$

8. $\dfrac{6}{7}$
    $-\dfrac{1}{7}$

9. $\dfrac{11}{16}$
    $-\dfrac{3}{16}$

10. $\dfrac{5}{8}$
    $-\dfrac{3}{8}$

11. $\dfrac{10}{17} - \dfrac{3}{17} =$ _____

12. $\dfrac{9}{20} - \dfrac{3}{20} =$ _____

13. $\dfrac{13}{14} - \dfrac{7}{14} =$ _____

Compare. Write >, <, or =.

14. $\dfrac{5}{6} - \dfrac{4}{6}$ ___ $\dfrac{3}{6} - \dfrac{1}{6}$

15. $\dfrac{8}{13} - \dfrac{3}{13}$ ___ $\dfrac{7}{13} - \dfrac{5}{13}$

16. $\dfrac{10}{11} - \dfrac{6}{11}$ ___ $\dfrac{8}{11} - \dfrac{4}{11}$

Problem Solving

17. A salad contains $\dfrac{7}{8}$ cup of chopped celery and $\dfrac{3}{8}$ cup of olives.

    How much more celery than olives is in the salad? _____

18. At the first shop, $\dfrac{3}{10}$ of the tulips are red and $\dfrac{1}{10}$ are purple.

    What fraction are neither red nor purple? _____

Spiral Review

Find the fraction of the number.

19. $\dfrac{1}{10}$ of 50 = _____

20. $\dfrac{1}{3}$ of 9 = _____

21. $\dfrac{1}{6}$ of 54 = _____

Name_____

12·3 **Problem Solving: Reading for Math**

Choose an Operation

Solve. Tell how you chose the operation.

1. Jerome buys 16 pounds of chopped meat for a cookout but uses only 10 pounds. What fraction of the original supply is left?

_____

_____

2. Daniela uses $\frac{3}{8}$ of one package of cheese and $\frac{5}{8}$ of another to make sandwiches. How much does she use in all?

_____

_____

Use data from the recipe for problems 3–5.

Snack Mix—Serves 1

$\frac{3}{8}$ cup salted peanuts	
$\frac{3}{8}$ cup salted pretzel sticks	
$\frac{1}{8}$ cup chocolate chips	
$\frac{1}{8}$ cup raisins	

3. What is the total amount of all the ingredients in the recipe?

_____

4. What is the total amount of salty food in the recipe?

_____

5. How would you change the recipe to make it for two? Write the amounts of the ingredients for that recipe.

_____

_____

Spiral Review

6. $\frac{1}{8}$ of 16 = _____

7. $\frac{1}{8}$ of 8 = _____

8. $\frac{1}{10}$ of 10 = _____

Name_____

 **12·4** # Explore Adding Fractions with Unlike Denominators

Add. Write the answer in simplest form.

1. $\frac{3}{4} + \frac{5}{8} =$ _____

2. $\frac{2}{3} + \frac{5}{12} =$ _____

3. $\frac{3}{5} + \frac{4}{5} =$ _____

4. $\frac{3}{5} + \frac{7}{10} =$ _____

5. $\frac{2}{5} + \frac{1}{10} =$ _____

6. $\frac{1}{7} + \frac{1}{14} =$ _____

7. $\frac{1}{16} + \frac{1}{8} =$ _____

8. $\frac{1}{16} + \frac{1}{4} =$ _____

9. $\frac{1}{16} + \frac{1}{2} =$ _____

10. $\frac{1}{6} + \frac{7}{12} =$ _____

11. $\frac{1}{2} + \frac{1}{3} + \frac{1}{6} =$ _____

12. $\frac{1}{2} + \frac{1}{4} + \frac{1}{8} =$ _____

13. $\frac{1}{4} + \frac{1}{12} =$ _____

14. $\frac{5}{6} + \frac{1}{3} =$ _____

15. $\frac{1}{5} + \frac{1}{10} =$ _____

16. $\frac{4}{5} + \frac{3}{10} =$ _____

17. $\frac{3}{12} + \frac{1}{6} =$ _____

18. $\frac{3}{5} + \frac{2}{5} + \frac{1}{10} =$ _____

19. $\frac{1}{3} + \frac{2}{3} + \frac{2}{9} =$ _____

20. $\frac{1}{7} + \frac{1}{49} =$ _____

21. $\frac{7}{12} + \frac{5}{6} =$ _____

Solve.

22. When serving juice to her family, Kate uses $\frac{1}{4}$ of one container and $\frac{7}{16}$ of another. What fraction of a container does she use in all?

    _____

23. Of the people at a cookout, $\frac{4}{9}$ eat hamburgers and $\frac{1}{3}$ eat hot dogs. What fraction eat either hamburgers or hotdogs?

    _____

Spiral Review

Write in simplest form.

24. $\frac{4}{22}$ _____

25. $1\frac{6}{9}$ _____

26. $\frac{4}{5}$ _____

Name_____

Add. Write each sum in simplest form.

1. $\dfrac{2}{3}$
    $+\dfrac{5}{9}$

2. $\dfrac{4}{7}$
    $+\dfrac{3}{14}$

3. $\dfrac{7}{12}$
    $+\dfrac{3}{4}$

4. $\dfrac{5}{6}$
    $+\dfrac{11}{12}$

5. $\dfrac{2}{9}$
    $+\dfrac{1}{3}$

6. $\dfrac{3}{4}+\dfrac{7}{8}=$ _____

7. $\dfrac{1}{6}+\dfrac{5}{12}=$ _____

8. $\dfrac{1}{5}+\dfrac{7}{15}=$ _____

9. $\dfrac{1}{9}+\dfrac{2}{3}=$ _____

10. $\dfrac{1}{2}+\dfrac{5}{6}+\dfrac{2}{3}=$ _____

11. $\dfrac{2}{5}+\dfrac{1}{3}+\dfrac{7}{15}=$ _____

Compare. Write >, <, or =.

12. $\dfrac{1}{2}+\dfrac{3}{12}$ —— $\dfrac{1}{6}+\dfrac{1}{4}$

13. $\dfrac{1}{4}+\dfrac{3}{8}$ —— $\dfrac{1}{8}+\dfrac{1}{2}$

14. $\dfrac{1}{2}+\dfrac{3}{16}$ —— $\dfrac{1}{4}+\dfrac{5}{8}$

Problem Solving

15. Jayne's salad is made up of 3 cherry tomatoes, 7 cucumber slices, and 4 strips of green pepper. What fraction of the total items is each of the different kinds of vegetables?

_____

16. If Jayne leaves on her plate 1 cherry tomato, 2 cucumber slices, and 1 strip of green pepper, what fraction of the total items did she eat?

_____

Spiral Review

Use mental math.

17. $16 \times 50 =$ _____

18. $16 \times 500 =$ _____

19. $16 \times 5,000 =$ _____

Name_____

12·6 ▶ **Problem Solving: Strategy**

Solve a Simpler Problem

Solve.

1. Ellen buys an 8-pound pumkin for $0.59 a pound and 5 pounds of red potatoes for $0.68 a pound. Clarissa buys 3 pounds of onions for $0.69 a pound and a 6-pound bag of baking potatoes for $0.72 a pound. Who spends more money? How much more? _____

2. Recipe A uses these vegetables: $\frac{3}{4}$ cup of tomatoes, $\frac{1}{2}$ cup of green pepper, $\frac{1}{3}$ cup of hot peppers. Recipe B uses these vegetables: $\frac{7}{8}$ cup of tomatoes, $\frac{1}{3}$ cup of green pepper, $\frac{1}{8}$ cup of hot peppers, $\frac{1}{3}$ cup of cucumbers. Which recipe uses more vegetables? How much more? _____

Mixed Strategy Review

5. LaToya buys 2 pounds of potatoes at $0.59 a pound and 4 lemons at $0.89 each. Her change is $15.26. How much money did she give the clerk? _____

6. There are 24 pieces of fruit on a tray. There are 4 times as many apples as there are kiwis, and 6 more oranges than kiwis. How many kiwis are on the tray? _____

7. A recipe calls for $\frac{2}{3}$ cup of rice. How much rice will you use if you want to double the recipe? Triple it? _____

Spiral Review

Write >, <, or =.

8. $\frac{15}{16}$ ____ $\frac{3}{4}$

9. $\frac{15}{16}$ ____ $\frac{7}{8}$

10. $\frac{15}{16}$ ____ $\frac{30}{32}$

Name_____

 **12-7** **Explore Subtracting Fractions with Unlike Denominators**

Subtract. Write each answer in simplest form.

1. $\dfrac{4}{5} - \dfrac{3}{10} = $ _____   2. $\dfrac{5}{7} - \dfrac{1}{14} = $ _____   3. $\dfrac{3}{4} - \dfrac{5}{8} = $ _____   4. $\dfrac{7}{10} - \dfrac{3}{5} = $ _____

5. $\dfrac{7}{8} - \dfrac{1}{4} = $ _____   6. $\dfrac{11}{12} - \dfrac{1}{2} = $ _____   7. $\dfrac{7}{16} - \dfrac{1}{8} = $ _____   8. $\dfrac{1}{3} - \dfrac{1}{9} = $ _____

9. $\dfrac{11}{16} - \dfrac{1}{2} = $ _____   10. $\dfrac{5}{6} - \dfrac{7}{12} = $ _____   11. $\dfrac{1}{2} - \dfrac{1}{16} = $ _____   12. $\dfrac{9}{10} - \dfrac{1}{5} = $ _____

13. $\dfrac{11}{15} - \dfrac{2}{5} = $ _____   14. $\dfrac{7}{20} - \dfrac{1}{10} = $ _____   15. $\dfrac{2}{5} - \dfrac{1}{10} = $ _____   16. $\dfrac{7}{5} - \dfrac{3}{10} = $ _____

17. $\dfrac{7}{9} - \dfrac{2}{3} = $ _____   18. $\dfrac{13}{15} - \dfrac{2}{5} = $ _____

Solve.

19. A fruit punch recipe calls for $\dfrac{7}{8}$ of a gallon of orange juice. Elizabeth has

$\dfrac{3}{4}$ of a gallon. What fraction of a gallon does she still need? _____

20. Yusef has a box containing $\dfrac{15}{16}$ of a pound of breakfast cereal. He pours $\dfrac{1}{2}$

pound of the cereal into breakfast bowls. How much is still in the box?

_____

Spiral Review

21. 546 − 54 = _____   22. 564 − 45 = _____

23. 645 − 64 = _____   24. 654 − 46 = _____

12·8 Subtract Fractions with Unlike Denominators

1. $\dfrac{2}{3}$
   $-\dfrac{3}{8}$

2. $\dfrac{4}{7}$
   $-\dfrac{3}{21}$

3. $\dfrac{9}{10}$
   $-\dfrac{3}{5}$

4. $\dfrac{5}{6}$
   $-\dfrac{3}{4}$

5. $\dfrac{7}{12}$
   $-\dfrac{1}{6}$

6. $\dfrac{3}{4} - \dfrac{1}{8} =$ _____

7. $\dfrac{1}{6} - \dfrac{1}{8} =$ _____

8. $\dfrac{1}{3} - \dfrac{1}{9} =$ _____

Problem Solving

9. Janette buys $\dfrac{1}{2}$ pound of powdered ginger and uses $\dfrac{1}{16}$ pound. How much does she have left? Write your answer as a fraction of a pound and in ounces.

   _____

Use Data from this recipe for problems 11–12

10. How much more water than rice is in the pudding?

    _____

11. To double the recipe, how much of each ingredient is needed?

    _____

    _____

    _____

Rice Pudding—Serves 4	
$\dfrac{1}{3}$ cup rice	1 cup milk
$\dfrac{3}{4}$ cup water	$\dfrac{1}{4}$ tsp. vanilla
1 egg	$\dfrac{1}{8}$ tsp. salt
$\dfrac{1}{4}$ cup sugar	$\dfrac{1}{8}$ tsp. nutmeg
$\dfrac{1}{4}$ cup raisins	

Spiral Review

Use mental math.

12. $900 \div 15 =$ _____

13. $9,000 \div 15 =$ _____

14. $9,000 \div 150 =$ _____

Name_____

12·9 ▶ Properties of Fractions

Use properties to find each missing number.

1. $\frac{1}{4} + \left(\frac{3}{4} + \frac{5}{16}\right) = \left(\frac{1}{4} + \underline{\quad}\right) + \frac{5}{16}$

2. $\frac{3}{7} + \underline{\quad} = \frac{3}{7}$

3. $\frac{1}{5} + \frac{3}{5} = \underline{\quad} + \frac{1}{5}$

4. $\left(\frac{7}{9} + \underline{\quad}\right) + \frac{1}{9} = \frac{7}{9} + \left(\frac{1}{3} + \frac{1}{9}\right)$

Add. Then use the property to write a different number sentence.

5. $\frac{3}{5} + \frac{3}{10}$ Commutative _____

6. $\frac{6}{7} + \frac{6}{7}$ Identity _____

7. $\frac{8}{9} + \left(\frac{5}{9} + \frac{1}{9}\right)$ Associative _____

Problem Solving

Use data from the pictograph for problems 8–10.

8. The cereal with the most votes got how many more votes than the cereal with the least votes?

   _____

9. What fraction of the total number prefers Corny Flakes?

   _____

10. Which combinations of cereals got as many votes as Oat Squares?

    _____

Students' Favorite Cereals

Corny Flakes	🥣 🥣 🥣 🥣
Honey Bs	🥣 🥣 🥣
Oat Squares	🥣 🥣 🥣 🥣 🥣 🥣
Raisin Grain	🥣
Cheery Yums	🥣 🥣

Each 🥣 represents 5 votes.

Spiral Review

Use mental math.

11. $300,000 \div 50 = \underline{\quad}$

12. $30,000 \div 500 = \underline{\quad}$

13. $30,000 + 5,000 = \underline{\quad}$

14. $3,000 \times 50 = \underline{\quad}$

 **13·1** # Explore Fractions and Decimals

Write a fraction and a decimal for each shaded part. Then write the fraction in simplest form.

1. 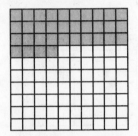 _____

2. 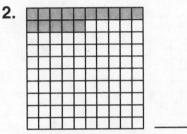 _____

3. 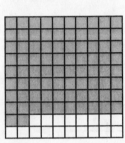 _____

4. 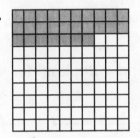 _____

Write each as a decimal.

5. $\frac{3}{10}$ _____

6. $\frac{45}{100}$ _____

7. $\frac{2}{5}$ _____

8. $\frac{1}{2}$ _____

9. $\frac{99}{100}$ _____

10. $\frac{44}{100}$ _____

Solve.

11. One rainy day 0.6 inches of rain fell. Write this number as a fraction. _____

12. In a box of 24 crayons, 18 are broken. What fraction of the crayons is broken? Write this number as a decimal. _____

Spiral Review

Write in simplest form.

13. $\frac{45}{100}$ _____

14. $\frac{44}{100}$ _____

15. $\frac{30}{100}$ _____

Name_____

 **13·2** **Tenths and Hundredths**

Write a fraction and a decimal for each part that is shaded. Then write the fraction in simplest form.

1.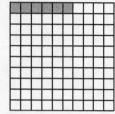

2.

3.

_____   _____   _____

Write a fraction and a decimal for each point. Tell if it is closest to 0, $\frac{1}{2}$, or 1.

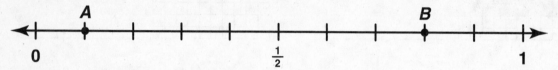

4. A _____   5. B _____

Problem Solving

6. Of 100 people watching the baseball game, 48 are wearing red caps. Write a fraction and a decimal for the number of people wearing red caps.   _____

7. Of 10 fish in the tank, 8 are goldfish. Write a fraction and a decimal for the number of fish that are not goldfish.   _____

Spiral Review

Write in simplest form.

8. $\frac{14}{18}$ _____   9. $\frac{12}{16}$ _____   10. $\frac{10}{14}$ _____

Name_____

 **13·3** **Thousandths**

Write each as a decimal.

1. $\dfrac{37}{1,000}$ _____

2. $\dfrac{56}{1,000}$ _____

3. $\dfrac{101}{1,000}$ _____

4. $\dfrac{96}{1,000}$ _____

5. $\dfrac{555}{1,000}$ _____

6. $\dfrac{600}{1,000}$ _____

7. $\dfrac{94}{1,000}$ _____

8. $\dfrac{2}{1,000}$ _____

9. $\dfrac{3,922}{10,000}$ _____

10. $\dfrac{4,217}{10,000}$ _____

11. three hundred four thousandths _____

12. one hundred six thousandths _____

13. seventy-seven thousandths _____

14. one hundred thousandths _____

Problem Solving

15. A baseball team won 0.620 of its games. Write the word name for this decimal.

_____

16. A Major League baseball game is at least 9 innings long. If a team completes 162 games a season, what is the minimum number of innings it plays?

_____

Spiral Review

Add. Write each sum in simplest form.

17. $\dfrac{1}{8} + \dfrac{3}{4} =$ ____

18. $\dfrac{3}{10} + \dfrac{3}{10} =$ ____

19. $\dfrac{1}{8} + \dfrac{3}{8} =$ ____

20. $\dfrac{9}{10} + \dfrac{9}{10} =$ ____

Name_____

Choose a Representation

Choose a representation and solve.

1. Commuters were asked how far they lived from their train station. Of those who answered, 0.2 said they lived less than 1 mile away; $\frac{1}{2}$ said they lived more than 2 miles away. How many lived between 1 and 2 miles from the station? Did a greater number live more or less than 2 miles from the station?

   _____

2. Commuters who live more than a mile from their station were asked how they got there. Of those who answered, 0.10 said they walked, 0.10 said they were driven by someone else, and 0.05 said they took a bus or taxi. The rest drove their own cars to the station. How many was that? Was it greater than all the other amounts put together?

   _____

Use data from the table for problems 3–6.

How People Get to Work

3. What fraction represents the share of people who drive alone? _____

4. What decimal represents the share of people who take either the train or the subway? _____

5. Donna says that $\frac{1}{4}$ of city residents take the subway. Is her statement reasonable? _____

Method	Share of workers
Drive alone	0.30
Carpool	0.10
Train	0.20
Subway	0.25
Bus	0.10
Walk	0.05

6. What fraction of those surveyed walk to work? _____

Spiral Review

Write >, <, or =.

7. $\frac{1}{20}$ _____ $\frac{1}{10}$

8. $\frac{2}{10}$ _____ $\frac{1}{5}$

9. $\frac{3}{10}$ _____ $\frac{1}{4}$

Name_____

Write a mixed number and a decimal to tell how much is shaded.

1.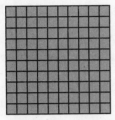

   _____

2.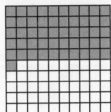

   _____

Write the decimal.

3. $7\dfrac{7}{10}$ _____

4. $2\dfrac{36}{1,000}$ _____

5. $5\dfrac{27}{100}$ _____

6. forty-two and twenty-four one thousandths _____

7. two hundred and sixty-two hundredths _____

Problem Solving

Use data from the chart for problems 8–9.

8. The population in 1970 was about how many times larger than it was in 1950?

   _____

9. In which year was the population between 6 and 7 times what it was in 1950?

   _____

Las Vegas Population, 1950–1990	
1950	24,624
1970	125,787
1980	164,674
1990	258,204

Spiral Review

Estimate.

10. 18×36 _____

11. 23×57 _____

12. 59×82 _____

 **13·6** **Compare and Order Decimals**

Compare. Write >, <, or =.

1. 0.53 _____ 0.053   2. 1.94 _____ 1.949   3. 12.5 _____ 12.500

4. 5.36 _____ 5.39   5. 0.3 _____ 0.03   6. 4.750 _____ 4.7

Write in order from greatest to least.

7. 0.343, 0.434, 0.35 _____

8. 98.9, 9.79, 89.78 _____

9. 11.063, 11.12, 11.014 _____

Write in order from least to greatest.

10. 1.714, 3.141, 1.317 _____

11. 4.045, 4.504, 4.5 _____

12. 3.933, 3.936, 3.693 _____

Problem Solving

Use data from the chart for problems 13–14.

13. Which place is more than 6.0 but less than 6.5 miles from Alison's house?

_____

14. Write in order from greatest to least the three places that are farthest from Alison's house.

_____

Miles from Alison's House	
Airport	8.6
Art museum	5.8
Bus station	6.2
City hall	6.6
County courthouse	6.7
Train station	5.5

Spiral Review

Write each as a decimal.

15. $\frac{3}{5}$ _____   16. $\frac{57}{100}$ _____   17. $\frac{7}{10}$ _____

13·7 Problem Solving: Strategy

Draw a Diagram

Draw a diagram to solve.

1. Petra lives 1.2 miles east of Warren Elementary School. Pierce Middle School is 1.4 miles west of her house. MacArthur High School is 0.8 miles east of Warren. Which school is closest to her house? Which is the farthest from her house?

_____

2. The aquarium is 0.5 kilometers south of the museum. Earl lives 2.35 kilometers north of the museum. Tom lives 2.14 kilometers south of the aquarium. Carla lives 1 kilometer south of Tom. Who lives closest to the museum? Who lives farthest from the museum?

_____

3. When Ronette leaves her office, she walks up the hall 4 offices, where she meets Dennis. She and he walk back down the hall 9 offices. Then Dennis returns to his own office, which is 3 offices back up the hall. How many offices away is Ronette's office from Dennis'?

_____

Mixed Strategy Review

5. Lynn needs to earn $3.65. She earns 3 rolls of 25 nickels each. Does she have enough money?   _____

6. Elizabeth goes downtown with her mother to their favorite bead store. They buy 5 large blue-and-green beads for $2.98 each, 26 smaller green beads for $0.40 each, 25 gold beads for $0.50 each, and a clasp for $1.25. How much do they spend in all?   _____

Spiral Review

7. $56.32 + $22.29 = _____

8. $56.32 − $22.29 = _____

9. $56.32 × 22 = _____

10. $56.32 ÷ 22 = _____

13·8 Round Decimals

Round to the nearest whole number.

1. 45.29 _____   2. 13.34 _____   3. 29.79 _____

4. 34.61 _____   5. 25.22 _____   6. 15.82 _____

Round to the nearest tenth.

7. 6.17 _____   8. 22.25 _____   9. 10.33 _____

10. 14.18 _____   11. 51.32 _____   12. 76.497 _____

Round to the nearest hundredth.

13. 7.551 _____   14. 23.435 _____ 15. 10.022 _____

16. 17.674 _____ 17. 9.909 _____   18. 38.3326 _____

Problem Solving

19. Dawn's mother's car averages 20.3 miles to the gallon of gasoline in the city. What is this amount rounded to the nearest whole number?

_____

20. Here are the average miles to the gallon of some of the family cars on the block: Wesson, 20.6; Barker, 23.4; Shore, 22.9; Milano, 21.6. Whose car gets the most miles to the gallon? The fewest?

_____

Spiral Review

Write each as a decimal.

21. $\frac{59}{100}$ _____   22. $\frac{66}{1,000}$ _____

23. $14\frac{1}{10}$ _____   24. $23\frac{4}{5}$ _____

14·1 Explore Adding Decimals

Find each sum.

1. 0.3 + 0.4 = _____   **2.** 0.3 + 0.8 = _____

3. 0.8 + 0.15 = _____   **4.** 0.85 + 0.85 = _____

5. 1.23 + 1.4 = _____   **6.** 1.49 + 1.36 = _____

7. 1.20 + 1.88 = _____   **8.** 1.79 + 0.05 = _____

9. 1.83 + 0.69 = _____   **10.** 1.25 + 1.47 = _____

11. 2.25 + 1.03 = _____   **12.** 2.87 + 0.45 = _____

13. 2.51 + 1.29 = _____   **14.** 2.7 + 1.4 = _____

15. 2.71 + 1.13 = _____   **16.** 2.2 + 1.68 = _____

Solve.

17. At the farmer's market, Frances buys tomatoes for $2.49, green peppers for $1.65, and some flowers for $4.99. How much does she spend in all?

_____

18. The bus that takes Philip to school travels 2.6 miles along Route 3, 4.65 miles along 31st Street, and 1.2 miles along Rugby Road. How far does it travel in all?

_____

Spiral Review

19. $\dfrac{5}{9} + \dfrac{5}{9} =$ _____   **20.** $\dfrac{4}{13} + \dfrac{7}{13} =$ _____   **21.** $\dfrac{5}{14} + \dfrac{3}{14} =$ _____

Name_____

14·2 Add Decimals

Add.

1. 5.6
    + 2.2

2. 0.67
    + 2.23

3. 3.55
    + 1.81

4. 15.6
    + 3.8

5. 4.563
    + 2.212

6. 9.015
    + 3.343

7. 22.27
    14.26
    + 3.59

8. 7.163
    2.558
    + 0.27

9. 68.213
    + 0.421

10. 23.16
     + 4.92

11. 0.364
     + 6.421

12. 8.267
     + 5.895

13. 0.89 + 0.4 + 0.288 = _____

14. 1.546 + 0.9 + 0.87 = _____

Problem Solving

15. Lupe sends 3 packages to her grandmother. The first weighs
    3.45 lb, the second weighs 2.67 lb, and the third weighs 1.95 lb.
    How much do they weigh in all?   _____

16. Shipping costs for packages are $3.00 for the first pound
    and $2.80 for each pound or part of a pound after that. How
    much does it cost to ship a package that weighs 3.68 lb? _____

17. Web watches his dad weigh cartons before loading them onto
    a cargo plane. They weigh 51.25 lb, 40.67 lb, and 34.45 lb.
    How much do they weigh in all?   _____

Spiral Review

18. 37,500 ÷ 80 = _____

19. 24,900 ÷ 30 = _____

20. 50,000 ÷ 12 = _____

21. 27,000 ÷ 15 = _____

Name_____

 **14·3** **Estimate Sums**

Estimate. Round to the nearest whole number.

1. 6.9 + 3.6 _____

2. 5.92 + 5.41 _____

3. 11.27 + 8.12 _____

4. $13.86 + $11.04 _____

Add. Check your answers.

5. 7.67 + 2.23 = _____

6. 11.35 + 2.64 = _____

7. $41.33 + $7.65 = _____

8. 4.9 + 5.6 + 2. 9 = _____

9. 4.21 + 2.68 + 1.12 = _____

10. $3.37 + $0.93 + $1.30 = _____

Problem Solving

Use data from the table for problems 11–12.

Distances Between Towns (in miles)

	Arrowville	Barnard	Gaston	Hallowell
Arrowville	—	6.8	7.2	15.8
Barnard	6.8	—	11.1	9.0
Gaston	7.2	11.1	—	23.2
Hallowell	15.8	9.0	23.2	—

11. Esther's mother drives her from their home in Barnard to Arrowville. Next they go from Arrowville to Gaston. From Gaston they return home. About how many miles have they driven? _____

12. Which two towns are the farthest apart? Which are closest?

_____

Spiral Review

13. 20 cm = _____ dm

14. 20 dm = _____ m

15. 2,000 m = _____ km

16. 2,000 m = _____ cm

Name_____

Choose an Operation

Solve. Tell how you chose the operation.

1. Ben's father is a bus driver. Yesterday he drove 122.4 miles. Today he drove 15.3 miles farther than yesterday. How far did he drive today?

   _____

2. Elroy's father is a bus driver too. Last week he drove 620.8 miles. If he drives the same amount this week, how many miles will he drive in all?

   _____

Use data from the table for problems 3–6.

County Bus Route Distances Between Stops (Miles)	
Bern Depot to Kane Hospital	2.3
Kane Hospital to Town Hall	1.2
Town Hall to Museum	2.2
Museum to Oak Square	2.5
Oak Square to Central High	0.8
Central High to Kings Mall	2.6
Kings Mall to Penn Depot	3.7

3. Regina's mother is a nurse at Kane Hospital. After work today, she will ride the bus to a PTA meeting at Central High. How far will she ride on the bus? _____

4. Carly's mother also works at Kane Hospital. Today after work she will ride the bus to Penn Depot. How much farther will she travel than Regina's mother? _____

5. After work at the Museum, Keisha's father has to take the bus back to Bern Depot. How many miles will he travel? _____

6. If the bus route is extended to Blake State Park, 4.3 miles from the Penn Depot, how long will the total route be? _____

Spiral Review

Change each fraction to a decimal.

7. $\dfrac{3}{1,000}$ _____   8. $\dfrac{27}{100}$ _____   9. $\dfrac{151}{1,000}$ _____

14·5 Explore Subtracting Decimals

Find each difference.

1. 1.88 − 1.02 = _____

2. 1.9 − 1.52 = _____

3. 0.99 − 0.8 = _____

4. 0.82 − 0.47 = _____

5. 0.8 − 0.3 = _____

6. 1.6 − 0.08 = _____

7. 1.75 − 0.25 = _____

8. 1.33 − 0.67 = _____

9. 1.00 − 0.07 = _____

10. 1.34 − 0.99 = _____

11. 1.5 − 0.87 = _____

12. 1.46 − 0.21 = _____

13. 1.60 − 0.90 = _____

14. 0.20 − 0.03 = _____

15. 1.7 − 1.66 = _____

16. 1.7 − 0.66 = _____

17. 1.99 − 0.98 = _____

18. 1.35 − 0.36 = _____

19. 1.08 − 0.89 = _____

20. 1.66 − 0.2 = _____

Solve.

21. Mindy's sailboat is traveling at 4.75 miles per hour. Warren's motorboat is traveling at 8.5 miles per hour. How much faster is the motorboat traveling?

_____

22. The San Jose Sports Truck is 4.28 meters long. The Urbana sedan is 3.85 meters long. How much longer is the Sports Truck? _____

Spiral Review

23. 3.43 + 0.99 = _____

24. 2.16 + 0.3 + 0.9 = _____

25. 11.15 + 1.02 + 3.53 = _____

26. $12.90 + $1.66 + $3.45 = _____

Name_____

14·6 Subtract Decimals

Subtract. Check each answer.

1.   1.67
    − .93

2.   0.701
    − 0.098

3.   10.657
    − 3.899

4.   5.87
    − 2.33

5.   4.555
    − 0.089

6.   11.101
    − 0.884

7.   16.302
    − 0.629

8.   4.21
    − 3.89

9.   10.353
    − 4.102

10. 15.010 − 0.003 = _____

11. 8.090 − 4.596 = _____

Problem Solving

Use data from the table for problems 12–15.

12. Of the cars that entered the race, $\frac{5}{7}$ were from the United States. How many were not from the United States? _____

An Early Coast-to-Coast Car Race

Kilometers	4,570
Number of cars entered	140
Winner's average speed	9.067 km/h
Winner's expenses	$3,558

13. The second-place car had an average speed of 8.821 kilometers per hour. How much faster on average was the winning car?

_____

14. One car only made 2,264 km. How much farther would it have had to go to finish the race? Did the car make it halfway? _____

15. The owners of the third-place car had $5,940 in expenses. How much less were the winner's expenses? _____

Spiral Review

16. 75,604 − 39,900 = _____

17. 39,900 ÷ 25 = _____

18. 75,604 × 3 = _____

19. 39,900 + 1,650 = _____

Name_____

14·7 Estimate Differences

Estimate. Round to the nearest whole number.

1. 19.45 – 6.8 _____

2. 11.05 – 5.6 _____

3. 7.73 – 4.89 _____

4. $35.60 – $12.99 _____

5. 13 – 8.8 _____

6. 15.44 – 12.06 _____

7. 72.06 – 3.51 _____

8. 9.29 – 8.78 _____

9. 10.71 – 3.06 _____

10. 14.3 – 9.7 _____

Subtract. Check your answers.

11. $90.34 – $35.99 = _____

12. $8.18 – $4.55 = _____

13. 14.90 – 4.78 = _____

14. 100.00 – 46.79 = _____

Problem Solving

15. If a 747 aircraft averages 517.9 miles per hour and a DC-9 averages 380.2 miles per hour, about how much faster does the 747 go on average?

_____

16. If the Concorde goes 1,113.8 miles per hour, about how much faster is that than the 747? _____

17. If the operating cost per hour of an airplane is $5,000 and the airplane averages 500 miles an hour, how much does it cost to fly a mile? _____

Spiral Review

18. $\dfrac{4}{9} + \dfrac{2}{3} =$ _____

19. $\dfrac{5}{12} + \dfrac{1}{6} =$ _____

20. $\dfrac{8}{15} + \dfrac{2}{5} =$ _____

21. $\dfrac{10}{17} + \dfrac{14}{17} =$ _____

Name_____

Solve Simpler Problems

Solve using a simpler problem.

1. A certain 747 aircraft uses 3,695 gallons of fuel an hour. How much fuel will it

   use on a 5-hour flight? _____

2. At the airport shop, 34 T-shirts were sold for $12.98 each, and 47 mugs were
   sold for $5.98 each. How much money did customers pay for T-shirts and
   mugs?

   _____

3. A steward earns $44.21 an hour and a flight attendant earns $39.78 an hour.

   How much do both workers earn for a 5-hour flight? _____

4. One plane travels at a speed of 539 miles an hour and another travels at a
   speed of 398 miles an hour. How much farther does the first plane travel in

   5 hours? _____

Mixed Strategy Review

5. The airport shop measures 30 feet by 20 feet. If its floor is made up of 12-

   inch-by-12 inch square tiles, how many tiles are there? _____

6. In the coach section of an airplane there are 2 seats on the left side, 4 in the
   middle section, and 2 on the right side. If the entire coach section holds 248

   passengers, how many rows of seats are there? _____

7. On a new airline, a plane ticket for a trip of less than 150 miles costs $40; a
   trip between 150 miles and 300 miles costs $80; a trip between 300 miles
   and 450 miles costs $120. If the price continues this way, how much will a

   flight of 1,100 miles cost? _____

Spiral Review

8. $567 ÷ 9 = _____

9. $45.25 × 20 = _____

10. $338.56 − $225.69 = _____

11. $459.32 + $612.30 = _____

Name_____

14·9 Use Properties to Add and Subtract

Add or subtract mentally.

1. 2.35 + 1.65 = _____

2. 5.89 + 4.21 = _____

3. 7.96 − 5.82 = _____

4. 8.99 − 5.51 = _____

5. 14.92 − 12.12 = _____

6. 1.6 + 3.9 + 4.4 = _____

7. 6.03 + 0.4 = _____

8. 16.02 + 4.34 = _____

9. 1.8 + 1.21 = _____

10. 9.4 − 0.7 = _____

11. 12.3 − 1.4 = _____

12. 4.06 + 7.32 = _____

Problem Solving

Use data from the chart for problems 13–16.

13. How much longer in miles is the English Channel tunnel than the Simplon Tunnel?

_____

14. How much shorter in kilometers is the Apennine tunnel than the Seikan tunnel?

_____

15. How long in miles are the Seikan and English Channel tunnels together?

_____

16. How long in kilometers are the three tunnels in the Alps together?

_____

Longest Railroad Tunnels*

Name	Length mi	km
Seikan (Japan)	33.5	53.9
English Channel	31.1	50.0
Simplon (Alps)	12.3	19.8
Apennine (Alps)	11.5	18.5
St. Gotthard (Alps)	9.3	15.0

*Time Almanac 2000

Spiral Review

17. 45.9 + 24.49 = _____

18. 54.4 − 49 = _____

19. 26.9 − 23.8 = _____

20. 80.8 − 49.97 = _____